A clear, comprehensive and unique book that provides rich guidance on b... interconnected web of life on and near farms can be managed to benefit p...
Professor Jules Pretty, Professor of Environment and Society
Director, Centre for Public and Policy Engagement (CPPE), University of Essex

John Bradshaw has produced in the pages of this book a resounding testament to the pioneering work of Alex Podolinsky and to the many farms, farmers and stewards of land that have taken up his thoroughly practical agroecological insights. Attested to by Bradshaw's own grasp of biodynamic methods and its benefits for land and people, the several case studies presented in the latter section of the book provide rich contextual examples of Podolinsky's lasting – and evolving – legacy.
Dr Jonathan M. Code, Royal Agricultural University, author of *Muck and Mind*

We have run our 10,000-acre wheat farm since 1986 using these practices, supplying clean food and being economically successful doing it. For anybody wanting to find a way of producing the most nutritious and chemical free food you can in your backyard or on a large-scale farm, this book has it all. John shows you how and why certain practices are done. This is a detailed instruction manual. Food and chemicals should never mix. This book will improve the health of all who follow its practices. Great book John.
Barry Edwards, Biodynamic grain producer, Victoria

A clear and accurate guide to Alex Podolinsky's Australian Demeter Biodynamic Method and a very useful resource for farmers.
Bridgette Olsen, Agricoltura Vivente, Italy, medicinal herb producer

This book that John has created is a wonderful resource and reference for new and experienced Biodynamic practitioners who are interested in enhancing their knowledge and understanding of the Australian Demeter Biodynamic way of farming and gardening. I have been a practising Biodynamic farmer for more than 35 years and I am still learning thanks to this book.
Mark Peterson, Biodynamic dairy farmer, Victoria

It covers both the theoretical and the operational aspects of Biodynamic practice in a thorough and very readable way. A very thorough, accessible and practical guide to Biodynamics for farmers and growers. I wish it had been available when we converted our orchard to Biodynamics 30 years ago!
Ian Burns, Biodynamic orchardist, Western Australia

I highly recommend this excellent book. It is well written and conveys in a practical way a deep understanding of the many elements important to success. The many stories from farmers provide a wealth of examples and experiences which are so valuable for the reader. I look forward to sharing it with my biodynamic colleagues in Denmark and also with newcomers to biodynamics.
Birthe Holt, Biodynamic vegetable grower and preparation maker, Denmark

Presents a very clear overview of the method, details the extensive scientific foundation of the method and provides a comprehensive guide to its practical application. As an Agricultural Engineer and Research Scientist, I can confidently state, based on my numerous farm scale trials, academic studies, and shared observations with the farming community that Biodynamics, properly applied, produces wholesome nutrient-rich plants grown unrestricted within the natural processes and are not limited as those plants grown using conventional agricultural methodologies. The resulting biologically dynamic soils increases the soil organic matter, humus, and structure to depth resulting in seasonally resilient productive soils: not depleted shallow soils as is the case resulting from high input conventional agricultural practice where ongoing financial sustainability of such practices is of major concern worldwide.
Dr John Russell PhD, former lecturer at La Trobe University

Alex Podolinsky is one of the major figures in the history of biodynamic agriculture. Working in the footsteps of Kolisko and Pfeiffer, he has helped practices and understanding evolve towards ever greater precision and efficiency. The colloidal quality of the preparations, the quality of their conservation and application, particularly the stirring, and respect for rigorous agronomic conditions have enabled biodynamic agriculture to reveal its extraordinary effectiveness in developing the structure and fertility of soils, the health of plants and the quality of agricultural products.

In his book, John Bradshaw presents the biodynamic method as developed in Australia. He gives many precise instructions and appeals to everyone's observation and common sense. This book contains the message that Alex Podolinsky passed on during his many travels and conferences, and which is still today a major source of inspiration for our work and that of many farmers and winegrowers around the world.
Vincent Masson, BioDynamie Services sarl

My 'elevator pitch' for Biodynamic farming? 'It's 95% good farming and 5% 'magic'… and we all need a bit of magic in our lives!' John Bradshaw's book focuses on the really good, practical farming that is real biodynamic farming, with insightful case studies from many farming sectors, countries and continents. He does not shy away from the 'magic' elements, or the preparations. These are explained clearly and practically with scientific backing as to how they work. I will be using this book as a resource in all my training work from now on.
Marina O'Connell, Director Apricot Centre, Huxhams Cross Farm, Devon,
author of *Designing Regenerative Food Systems*

John has the marvellous attitude to put aside his ego when presenting a farm or Alex's teaching. I remember Alex talking about pianists interpreting Bach, how they should serve Bach and not their ego. John is one of those amazing pianists who can serve Biodynamics, transmitting the knowledge, the emotions and the real experience. And he knows what it is all about, having practised it himself. But he never puts his experience over the farmers that he is listening to. Thank you John.
Saverio Petrilli, Biodynamic vigneron and adviser, Italy

Biodynamic Farming Handbook

Hawthorn Press

Biodynamic Farming Handbook © 2024 John Bradshaw

Published by Hawthorn Press, Hawthorn House,
1 Lansdown Lane, Stroud, Gloucestershire, GL5 1BJ, UK
Tel: (01453) 757040   Email: info@hawthornpress.com

Hawthorn Press
www.hawthornpress.com

Cover photo © Hyeri Choi: Field pea flowers, author's farm, Dumbalk, Victoria
Cover design and typesetting by Lucy Guenot
All photos are by John Bradshaw and Anna Su unless otherwise credited
Printed by Short Run Press, Exeter
Printed on environmentally friendly chlorine-free paper sourced from renewable forest stock

Every effort has been made to trace the ownership of all copyrighted material. If any omission has been made, please bring this to the publisher's attention so that proper acknowledgement may be given in future editions.

The views expressed in this book are not necessarily those of the publisher.

British Library Cataloguing in Publication Data applied for

ISBN 978-1-912480-93-7

Biodynamic Farming Handbook

Activating soil fertility for growing healthy food

John Bradshaw

Hawthorn Press

Dedication

Alex Podolinsky, 2017. Taken from the Podolinsky collection with kind permission.

I dedicate this book to the memory of Alex Podolinsky (1925–2019), who was the inspirational leader of the Australian Demeter Biodynamic agricultural movement. Working, from 1952, with creative and resourceful Australian farmers – men and women forged by an often hostile climate, ancient fragile soils, and lack of financial subsidisation – he developed the germinal ideas presented by Rudolf Steiner in 1924 into a fully fledged, science-based natural farming system. It was very much a developmental partnership, the farmers making a substantial contribution, particularly in the details of their own agricultural speciality.

So successful was Alex that by the early 1980s the Australian Demeter Biodynamic method was being applied to hundreds of thousands of acres of Australian farms and over the following decades it was widely adopted elsewhere, particularly in Europe and Asia. Alex's influence extended well beyond farming, encompassing education, the care of the intellectually disabled, architecture and much more.

Those who wish to read more about his life can read my article 'Alex Podolinsky – An Extraordinary Life' in *Biodynamic Growing*, 29 (2019).

https://biodynamicgrower.com/wp-content/uploads/2024/01/Alex-Podolinsky.pdf

Contents

Foreword

Alex Podolinsky was one of the most influential people in my life. He brought together disparate concepts and ideas in a way that made them accessible and practical for farmers, growers and viticulturalists. There is perhaps no better placed person to write about Alex's work in developing and spreading what has become known as the Australian Demeter Biodynamic method than John Bradshaw. John knew Alex for 45 years and has written about his work since the 1990s. John edited the *Biodynamic Growing* magazine, which kept the practitioners of Alex's Biodynamic method in a world-wide community of sharing practice.

Biodynamic practices are not always easily understandable. This can put people off before they have even started. The word Biodynamic comes from the combination of two Greek words *Bios* meaning life and *Dynamis* meaning force or energy. Biodynamics is simply bringing more life energy to the plants through the living soil, which spreads to those organisms that consume them, including us.

My thanks go out to all those Biodynamic farmers who inspired me through Alex's suggestions. The European group that followed Alex's approach developed in me a deeper understanding and appreciation of Biodynamic quality, particularly in the delicious, alive wines that we drank at the meetings. Alex often joined us at these meetings in Italy or France and shared the details of his approach.

Australia was where Alex came to farming after working for the Australian government to repay his and his wife's transport to Australia after World War II.[1] There was no place for Alex's Russian aristocratic ancestry in the egalitarian 'fair go' Australia. So Alex mucked in and developed his Australian Biodynamic farming method having been connected with the Biodynamic movement in Germany whilst growing up.

There has always been a particularly challenging nature to farming in Australia, where some of its ancient soils have had the nutrients washed out over millennia, others suffer from acute salinity or extreme climate, while others have a distinct fertility. Perhaps this is why the Biodynamic preparations had such a dramatically positive impact. In the 1970s, there was more Biodynamically certified land in Australia than organically certified land on the rest of the planet. Alex shot to fame in Australia when he featured in an ABC programme that highlighted his success. In my search for repairing and enhancing our soils, it was a chapter about Alex in the book *Secrets of the Soil* by Tompkins and Bird that grabbed my attention.

Alex from an early age had passions for music, particularly Bach, and for literature and philosophy. He would describe the interactions of life in the soil as akin to the music of an orchestra. Can you take one instrument or musician out of an orchestra and hope to understand how or what the orchestra creates? The dissection of the parts cannot truly mean understanding the whole organism. One must experience it as a whole. Martin Heidegger, a leading 20th Century philosopher, influenced Alex when at university aged 22. Or perhaps Alex influenced Heidegger, it is difficult to say. Heidegger is associated with phenomenology, the study of the appearance of things and the 'as a whole' or holistic world view.

Active perception[2] is one of Alex's greatest contributions. This is a process of learning that leads to seeing hitherto hidden natural processes, indeed to perceive these hidden living activities going on in the soil by seeing the natural living processes in the plants themselves. Through a series of observation exercises of handwritings, buildings and works of art, he describes the fluidity in life, the connectedness, and shows the importance of the whole.

Alex's description of Nature's organisation in plant feeding, which John details in this book, was one of the great formative 'light bulb' moments in my farming development. Understanding how plants are separately fed and watered in Nature, just as we are, was a pivotal moment for me. Natural plants drink what they need through transpiration and eat when they need from the living soil organisation, which Alex referred to as colloidal humus. Taking this further, one could liken the use of artificial fertilisers (or raw animal slurry), which work through the soil water to feed plants as akin to quenching the thirst of humans using only beer. There have been many recent scientific advances in understanding how plants, fungi and bacteria cooperate in this soil biosphere.

Alex, like many farmers before and since, has used the word humus to describe something more than simple soil organic matter. More specifically Alex used the term colloidal humus[3] to describe a highly developed, stable form of soil organic matter, a product of soil biota, that holds nutrients and water, buffers pH and provides soil structure. Defining the term colloidal humus is important because it focuses attention on the fact that not all organic matter is created equally.

This is something that became only too apparent for me a few years ago. Whilst visiting a farmer on a similar soil type to ours, I had a striking realisation, though now I wonder that I was surprised at all. The soil there was relatively high in soil organic matter as a result of returning crop residues to the soil and growing brassica-rich cover crops. What struck me was that the soil simply ran through my fingers like a sandy peat. 'What's wrong with that?' you might ask. Well, our Yatesbury Farm soil doesn't. Despite the similar soil types, their soils and ours were acting completely differently. It took me half an hour to resolve the conundrum, though Alex would not have been impressed by my slow thought process or slow active perception. There was little or no aggregation in this liquid soil. The soil particles and crop residue particles were not joined together. There were no soil glues (i.e. glycoproteins or glomalin related soil proteins) that have their origins mostly with the mycelium of arbuscular mycorrhizal fungi but also to a lesser extent from root exudates. That meant the soil was likely to be largely dead, especially to fungi which are the most important organisms in the soil aggregation process.

This is supported by research at Rothamsted by Professor Andrew Neal showing artificial fertilisers destroy soil organisms.[4] Fungicides sprayed on crops to protect them have an obvious negative effect on the soil fungi. The soil particles that are not aggregated are not stable, meaning the soil organic matter is more labile and readily degradable. So the term colloidal humus is used here as the shorthand for a living, energetic soil. We repaired and enhanced our soil through the use of the Biodynamic preparations which can be said to act primarily as a soil inoculant.

This handbook of the Australian Demeter Biodynamic Farming method cannot have come soon enough and will inform farmers transitioning to organic soil regeneration. It is timely, as it fits with the wider recognition that agriculture must change: must change to ensure the capacity to produce food for our children and grandchildren; must change to act in community with nature; and must change to provide healthy food, health which is infectious. As Lady Eve Balfour once said, 'The health of soil, plant, animal and man is one and indivisible'.

I owe an enormous debt of gratitude to Alex and his Biodynamic method. It has transformed our farm by repairing and enhancing our soils, but more than that, he has transformed me, to see or rather to actively perceive a conscious greater whole.

Richard Gantlett PhD
Yatesbury House Farm, Wiltshire.
Visiting Fellow in Agriculture,
University of Reading

FOOTNOTES FOR THE FOREWORD

1. A. Podolinsky, *Life: Contra Burocratismus.* (Alex Podolinsky, 2015).
2. A. Podolinsky, *Active Perception*, Bio-Dynamic Agricultural Association of Australia, 2008.
3. A. Podolinsky, *Bio-dynamic Agriculture Introductory Lectures*, (Gavemer Foundation Pub., 1985).
4. A. L. Neal, et al. Soil as an extended composite phenotype of the microbial metagenome. *Scientific Reports* 10, 10649, doi:10.1038/s41598-020-67631-0 (2020).

Acknowledgements

Thanks are due to: my sister, Dr Julie Bradshaw for proofreading and style guidance; Trevor Hatch of the Bio-dynamic Agricultural Association of Australia and Rod Turner of Biodynamic Growers Australia for checking the Biodynamic content and making many helpful suggestions; the Australian Demeter Biodynamic community of farmers and gardeners, worldwide, for their support of my Biodynamic publishing work since 2003 and their invaluable contributions to this volume through opening their farms to my visits, generously sharing their Biodynamic stories and practices and providing photographs; Martin Large, Publisher, Hawthorn Press, for his ongoing support and encouragement with the project; Claire Percival, Head of Production, for ensuring the smooth progress of the project; Lucy Guenot, typesetter, for her beautiful layout of the book; and, lastly, my wife Anna for her love, patience and endless support.

Preface

Never has there been a more pressing global need for practical guidance on growing healthy food in a way that regenerates soil fertility and cares for the environment. Over the past century, Rudolf Steiner's biodynamic agriculture has offered such an approach, demonstrating in all corners of the world that it is possible to grow healthy food in a way that enhances – rather than degrades – soils and the natural environment. Yet the take-up of biodynamic principles has, on most continents, been slow, not least because of the relative lack of readily available information and education. This has to change – globally we need more land farmed biodynamically, and more farmers to do the farming. This new book, by the distinguished Australian author John Bradshaw, makes a major contribution to achieving this through a highly readable and informative handbook that offers comprehensive guidance and insights relevant to those wishing to convert to biodynamic agriculture.

Taking his inspiration and learning from the legendary Alex Podolinsky (1925-2019), John Bradshaw has been at the heart of biodynamic farming and gardening in Australia for the last half century, as a farmer, gardener, author, trainer and Demeter certification inspector. He has previously published work on biodynamic practice and was editor of the *Biodynamic Growing* magazine from 2003-2019. He has dedicated the Handbook to Alex Podolinsky, recognising the pioneering work that Podolinsky and his group of farmers undertook in developing the distinctive 'Australian' approach to biodynamic practice that is described in the Handbook. While being developed in Australia, Bradshaw is at pains to point out that the approach has been successfully applied globally – there are many useful case studies in the book to illustrate this – and is thus not limited to its country of origin. Indeed, it seems to me that most of the approach – and handbook – will be utterly familiar to biodynamic farmers and gardeners – accepting that there are pioneers on most continents who, like Podolinsky, have developed particular insights that seek to interpret, contextualise and supplement Steiner's original lectures.

Leaving aside these differences, what appeals so much to me about this handbook is that its starting point is that everyone can become a biodynamic farmer or grower – and that they will want to once they have the opportunity. There is no mystic in this book; it offers practical insights and guidance that remind us of just how natural it is to farm biodynamically – and what most of us have lost over the last 100 years by not so doing. It concentrates – rightly – on the making and use of the Preparations – Steiner's unique contribution to truly regenerative agriculture – and the ways in which the Preparations form what Bradshaw refers to as the 'framework' upon which other biodynamic techniques can be applied. While not all authorities on, and practitioners of, biodynamics will agree about the use of the additional 'Australian' Preparation ('Prepared 500'), there is no doubt in my mind that the description of the Preparations is well handled, informative and absolutely core to the book. Equally informative are the case studies contained in the second half of the book. While most are from Australia there are European cases as well, from Germany, France, Italy and the UK. What they all have in common is a level of detail that allows the reader to gain a good insight into the multiple benefits of biodynamic farming. Why would any farmer not want to convert, having read these case studies?

Reflecting on my own experience, as a researcher, former Director of a biodynamic farm and a long-time biodynamic gardener, it will be such a gift to be able to recommend this handbook to anyone who asks about biodynamic practices – the enthusiasm with which it is written will surely capture them just as much as the text will inform them. And this is where I am sure that this handbook will prove invaluable, in inspiring those with an interest in biodynamics to take their first steps towards the practice. And this applies whether they are hobby gardeners or conventional/organic farmers looking for new ways in which to enhance soil activity as a means of improving the quality of the food that they grow. In this respect the chapter on transitioning to biodynamics really resonates with me. While it is surely important to encourage new entrants to farming and growing to choose biodynamics from the start, it is vital that

conventional farmers and growers feel that they have the opportunity and support to transition towards biodynamics – for it is here that we can make major strides towards achieving the full regenerative benefits of biodynamic agriculture. Encouraging transition has become an increasingly important theme within the 'alternative' farming movement in the UK; this handbook will surely speed this process.

To conclude, I congratulate John on writing such an exciting and excellent addition to the literature on biodynamic agriculture. It is surely a fitting tribute to the pioneering work of Alex Podolinsky that will, I am sure, catalyse reflection – and hopefully action – by farmers with an enquiring mind and a commitment to genuinely regenerative agriculture that produces truly healthy food.

Dr Neil Ravenscroft
Trustee, St Anthony's Trust; former Director
of Tablehurst Biodynamic Community Farm
and Honorary Research Fellow, School of Law,
Birkbeck College, University of London

Introduction

Biodynamics is an advanced organic method of agriculture – that is, it does not use artificial fertilisers or chemical pesticides, fungicides or herbicides. As with other organic methods, it aims to build soil humus levels and biological activity, enabling plants to feed and grow naturally. However, Biodynamics goes far beyond ordinary organic methods, using a range of powerful, naturally based preparations that greatly enhance soil biological activity and photosynthetic efficiency, resulting in healthier, more vibrant plants with exceptional flavour and keeping qualities, far beyond what is achievable through ordinary organic methods.

Biodynamics is in fact the first *modern* organic method of agriculture. For 10,000 years, agriculture was essentially organic. Chemical-based conventional agriculture only began in the mid 19th century; it gradually replaced traditional organic methods. Biodynamics began in 1924, with a series of eight lectures by Dr Rudolf Steiner (1861–1925), an Austrian philosopher and theosophist who had already made substantial contributions in fields as broad as education, science, medicine, intellectual disability, architecture, social reform and theology, to name a few. The modern organic movement began with the 1940 publication of Lord James Northbourne's book *Look to the Land*, in which the phrase 'organic farming' was first used. Northbourne based many of

his organic farming ideas on Rudolf Steiner's work, applying them on his family estate, and he hosted the first Biodynamic farming conference to be held in Britain.

Worldwide, Biodynamics is quite a broad church, which can cause some confusion. In different countries, it has developed in somewhat different ways, though the essentials, particularly the use of the Biodynamic preparations, have remained the same. Though Steiner insisted that his suggestions be thoroughly scientifically tested (and this was done with the preparations), some later developments in some countries have lacked scientific rigour. Steiner's lecture series provided the important fundamentals, but the overall method was still to be developed.

The Australian Demeter Biodynamic method was developed from 1952 by Alex Podolinsky (1925–2019), together with the farmers he inspired and taught, members of the Bio-dynamic Agricultural Association of Australia (BDAAA). Alex, assisted by the Australian farmers – whom he regarded as uniquely innovative and resourceful – developed the overall agricultural method as a practical and strictly science-based method. So successful was he that it was very widely adopted by professional Australian farmers and by the 1980s it was practised on hundreds of thousands of acres of arable land, proving equally successful in all Australian climatic regions, on all soil types and in all kinds of agricultural enterprises. Australian farmers receive no subsidies and must compete on a world market against often heavily subsidised overseas competitors. The fact that Biodynamics was so widely adopted in Australia is testament to its practicality and success as a farming method.

Based on figures published by Biodynamic authorities in Europe in the early 1980s, Australian Biodynamic farming constituted at that time over half of the world total of Biodynamic agriculture. The BDAAA was for many decades the largest natural farming body on earth by acreage of its members. This eventually ceased as vast cattle stations in Australia and other countries became organically certified, by virtue of the fact that they used no chemicals.

In the early 1990s, Alex was invited to visit Biodynamic farmers in Europe, since many of them felt they were not getting results as good as those in Australia. He continued visiting and teaching European farmers until 2019. Hundreds of farmers throughout Europe have adopted the Australian Demeter Biodynamic method, gaining similar excellent results to their Australian colleagues. Many farmers throughout Asia have also adopted the Australian Demeter method, with similar outstanding results.

My aim in this book is to present the Australian Demeter Biodynamic agricultural method as clearly as possible, outlining the scientific basis and practical application of the method, so that farmers in any country can see the possibilities for themselves, whether they are already farming organically and looking for a way to enhance the health and success of their operation, or farming conventionally but concerned about the health of their family, their use of chemicals, the future of their land and their long-term economic survival.

John Bradshaw

Part 1

Understanding Biodynamics

In Nature all things are in mutual interaction;
the one is always working on the other.

Rudolf Steiner

Chapter 1

Biodynamics: An Overview

Nature's organisation is a wonderful symphony of interrelatedness, intelligence and breathtaking design. To be successful, Biodynamic farmers need to develop a certain level of understanding of this complexity. We need to understand how plants are organised to grow naturally and we need to cooperate with this design, rather than apply a flawed agricultural theory to nature and try to make plants fit. The latter is a recipe for failure.

Biodynamic agriculture provides a clear, holistic overview of the plant, integrated with its overall environment of soil, water, air, light, warmth and cosmos. It provides a clear understanding of how plants grow within nature's organisation and, conversely, how plants are forced, in conventional agriculture (or poor organic practice), to grow outside nature's organisation, with all the problems that inevitably follow.

With an understanding of natural plant growth, and an overview of the total plant environment, Biodynamic farmers can produce food of the highest possible quality, from strong, resilient plants that grow relatively free of the pests and diseases that plague conventional growers and require toxic sprays; food that is beautifully flavoured, with excellent keeping quality; food that builds robust health in our families, our livestock and the consumers we supply.

We view the plant as part of a vibrant co-working of:

- Soil, with its teeming activity of microbes, fungi, enzymes, micro- and macro-fauna (including earthworms) and more. Plant roots actively structure the soil, in partnership with the soil life, and foster microbes and fungi in the rhizosphere (soil/root zone) through their secretions.
- Air in the soil and atmosphere (air is the most important component in soil).
- Water in the soil, atmosphere and plant.
- Sunlight and warmth.
- Influences from the moon and other cosmic bodies.
- Interactions with other plants growing nearby.

Soil

Soil is the most important basis of agriculture. Biodynamic techniques begin with soil development: a healthy, well-structured soil allows plants to grow as nature intends, vibrant and healthy. Everything follows from there. We have found the seven Biodynamic soil and compost preparations known as 500 and 502–507 to be the most powerful and effective substances in existence for soil development. Their use greatly enhances microbial activity, root development and humus formation, resulting in progressively deeper, darker, better-structured and more fertile soils. Preparation 501, the 'light spray', works on the upper plant, enhancing light metabolism. Many other Biodynamic practices are used to complement the work of the Biodynamic preparations. They will be introduced later. First, however, we need to understand the absolutely fundamental concept of natural plant feeding that forms the basis of Biodynamic agriculture.

Growing Plants within Nature's Organisation

Over 60 years ago, Alex Podolinsky came to understand the detailed inner workings of the plant from a uniquely new perspective. While lecturing to a group of farmers, a living, moving picture suddenly appeared between him and the audience, of the plant as a harmonious whole, complete with all its interactions with its environment.[1] It took Podolinsky a few years, all in the context of his own farming activities, to completely understand what he had visualised and be able to explain it to others. It now forms the basis of Australian Demeter Biodynamic practice and also underlies the (Australian) National Standard for Organic and Bio-Dynamic Produce, having been accepted as valid and fundamental by the Australian organic industry. This concept could well, in the future, be seen as the crucial pivotal point in agricultural history which restores agriculture to a valid developmental path after 180 years of problematic deviation following the discovery that plants could only absorb nutrients in solution, which led to the use of artificial fertilisers and the poisoning of our environment with the toxic chemical sprays needed to keep unnaturally fed plants functioning.

Let's start, then, with the concept of how plants feed as they are meant to feed – *within* nature's organisation, in contrast to the way plants feed *outside* nature's organisation. It is firstly important to understand the difference between colloidal humus – nature's preferred plant food – and organic matter.

Organic matter

Organic matter comprises anything that is alive, was alive or was produced by a living entity. The myriad components of the soil, including microbes, fungi, enzymes, co-enzymes, micro-fauna, macro-fauna and more, have one overriding function: to convert organic matter into colloidal humus.

Figure 1.1. Organic matter, decomposing

Figure 1.2. Colloidal humus

Colloidal humus

Colloidal humus is the final stage in the refinement of organic matter orchestrated by the soil life, particularly worms and microbes. A colloid lies between a solution and a suspension; examples are butter and jelly. Colloidal humus holds nutrients in a soluble form, available to plants' fine hair feeder roots that enter the colloid, but does not allow the nutrients to leave the colloid to dissolve in the soil water. Colloidal humus can hold up to 70% of its volume as water. It is moist, soft and malleable. Worm castings are one example of colloidal humus.

Growing within nature's organisation – natural plant nutrition

In Figure 1.3 we see a bean plant growing in a soil well supplied with humus particles (represented by the pink dots). Nutrients are contained mostly in the soil colloidal humus, in a soluble state, available to plant roots but held by the humus colloids. Because no water-soluble fertilisers have been used, the soil water is relatively pure. Plants have no independent warmth metabolism as animals do, and rely on sun warmth to tell them when to feed. When sun

Figure 1.3. Illustration by Jenny Bradshaw

Figure 1.4. Illustration by Jenny Bradshaw

4

warmth stimulates, the plant takes up nutrients from the soil humus, using fine hair feeder roots. Plants must transpire (evaporate water from the leaves) continuously just as we must breathe continuously, and the roots have to take water in from the soil to replace that lost from the leaves. Water uptake is essentially a separate process from nutrient uptake. Water is taken in by thicker more vertical roots and, in this situation, is relatively free of dissolved nutrients. The plant is metabolically balanced: nutrient uptake matches growth needs as dictated by the sun. The plant is upright, a healthy green colour and well furnished with beans. It is not attractive to insects and is resistant to disease.

Growing outside nature's organisation – unnatural plant nutrition

In Figure 1.4 we see a bean plant fertilised with water-soluble fertiliser (a situation equivalent to the application of raw manures or even poorly made composts). The nutrients, once dissolved, spread through the soil water (the pink colour of the soil in the picture is used to indicate the dissolved nutrients). As the plant takes up water (to replace that lost in transpiration), it must also take up nutrients (since they are dissolved in the water), independently of sun warmth – it must continuously feed. The role of the sun in governing feeding is diminished. The plant accumulates more nutrients than it can metabolically cope with and cells become overfull with elements such as nitrogen, phosphorus and potassium. The plant takes in more water to dilute these excess nutrients, but with the water come more nutrients. Cells become distended with mineral salts and water; the plant becomes overblown and darker green in colour. The guard cells of the stomata on the leaf also become distended, and so the exchange of gases and photosynthetic efficiency are affected. The plant is unbalanced and more susceptible to attack by insects and diseases. Flavour, nutritional value and keeping quality are impaired.

Key Biodynamic Techniques

Biodynamic soil development

Soil is the most important foundation of agriculture. The development of healthy, fertile, biologically active, well-structured soil is the highest priority for Biodynamic farmers. Biodynamic growers use a variety of techniques to develop well-structured, biologically active soils with healthy humus levels, so that plants can feed naturally. We never apply raw manures, or compost that has not been properly matured, to food crops for direct human consumption, since this would result in soluble nutrients spreading through the soil water, causing the same unnatural forced plant feeding as artificial fertilisers, with the same undesirable consequences. The most important soil development technique, and the one that most distinguishes Biodynamics from other organic methods, is the use of the Biodynamic preparations 500 and 502–507.

Preparation 500 is the most well known, being made by filling cow horns with fresh cow manure and burying them in the soil from autumn until spring. Preparation 500 is a very rich source of beneficial microbes. Liquefied, stirred in a special way for 1 hour and sprayed on moist, warm soil, it dramatically increases soil microbial activity, root growth and humus formation, building deeper, darker, better-structured soils. It also regulates soil pH, bringing it closer to neutral over time.

Preparations 502–507 are made from plant materials and are collectively called the 'compost preparations'. They are added to compost heaps to assist in the rapid and efficient transformation of organic material into colloidal humus. In the Australian Demeter Biodynamic method, they are also added to 500 to make 'prepared 500'. This process was developed by Podolinsky and thoroughly tested at the Bio-dynamic Research Institute (BDRI). It was proven considerably more effective in soil development than 'straight 500' (500 alone) and enables the six compost preparations to be regularly applied to the whole farm.

bringing enhanced flavour, sugar content and keeping qualities.

Warmth

Preparation 507 (valerian) has a unique warmth-bringing quality. It is very useful to spray it over plants when frosty conditions threaten crop damage. It has also been shown to greatly help plants to recover from severe storm damage.

Biodynamic compost

The compost preparations, 502–507, are used in compost heaps to produce 100% colloidal humus, which, turned into the soil or spread on pasture, is the ideal plant food. Good compost should be moist, dark and sweet-smelling and have a texture and kneadability similar to putty or modelling clay.

Figure 1.5. Spraying 500 – the soil fertility spray. Photo: Rob and Pauline Bryans.

Figure 1.6. Spraying 501 – the light spray. Photo: Saverio Petrilli.

Water/light

Preparation 501, made from finely powdered, clear quartz crystals, is sprayed as a fine mist above plant leaves. It greatly enhances light metabolism in the plant, overcoming wateriness, excessive lushness, and lack of light,

Biodynamic sheet composting

A form of in situ composting that involves careful pasture management, rotational grazing, harrowing, topping pasture, mowing of non-grazed pastures in horticulture, and green manuring.

Figure 1.7. Biodynamic compost preparations 502–507 are inserted in compost heaps to guide and assist the formation of colloidal humus and are added to 500 to make 'prepared 500'. Photo: Mark Peterson.

Cultivation

Biodynamic growers are extremely careful with soil cultivation, to preserve and develop soil structure. Cultivation is never done when soil is too wet or too dry. Tined implements are preferred, though a range of other implements may be used with due care. The Biodynamic Rehabilitator plough (see Chapter 16) is a useful implement for soil preparation, working in green manures and loosening subsoils and is used by many. It is not, however, recommended for pasture subsoiling.

Green manures

Biodynamic growers use green manures when required, to build high humus levels in the soil to feed subsequent crops. A wide variety of species is included in the seed mix. Preparation 500 is sprayed soon after sowing and after working in the green manure, to speed up its conversion into colloidal humus.

Figure 1.8. Green manuring.

Dynamic grass management

Rotational grazing is practised, with electric strip grazing where appropriate. This is an important factor in the ongoing development of soil structure. With cattle, pastures are harrowed after they have been grazed (provided the soil is not too wet), in order to spread manure evenly and thinly, allowing it to be quickly incorporated into the soil humus (this is not necessary when dung beetles are active).

Fertiliser inputs

Because of the highly developed soil life, open friable structure and high humus levels

Figure 1.9. Careful cultivation.

on Biodynamic farms, Biodynamic growers need fewer inputs than do ordinary organic or conventional growers on comparable soil types. If small amounts of soil amendments are needed, non-water soluble fertilisers are used, such as reactive phosphate rock or rock dusts.

Sowing by the moon and stars

Biodynamic growers follow a sowing chart developed as a result of research carried out over many years. By sowing at the correct time we can optimise results and also direct plants more towards leaf, root, flower or fruit. The sowing chart also provides the best times for transplanting, cutting hay, pruning and other activities.

Companion planting

Different plant species can influence each other's growth: some plants help others, some hinder others, and some combinations are mutually beneficial. Companion planting was used traditionally in many countries in the past. Biodynamic growers have greatly contributed to this pool of knowledge.

Figure 1.11. Sowing by the moon and zodiac. Biodynamic growers use a sowing chart developed from years of research. The timing of seed sowing can influence plants to grow more to root, leaf, flower or fruit. Optimal times can also be chosen for transplanting, cultivation and other tasks. This kind of practice is less feasible on broadacre farms, where perhaps a thousand acres of wheat will have to be sown quickly, over a week or so. Source: Universe (detail) by Haykuhi Khachatryan, 2015, with kind permission.

Figure 1.10. Dynamic pasture management and harrowing.

FOOTNOTE FOR CHAPTER 1

1. Not unlike the living pictures seen by Friedrich August Kekulé (1829–1896) which led him to the discovery of the chemical structure of benzene and other compounds.

Chapter 2

The Biodynamic Preparations

The eight Biodynamic preparations lie at the heart of Biodynamics. They form the essential framework to which the other Biodynamic techniques are added. Rudolf Steiner, in his 1924 lecture series, outlined the method of making the eight preparations: seven for developing biological activity, structure and balance in soils and one to enhance light metabolism in plants. In addition he recommended the use of *Equisetum arvense* (horsetail) tea[1] as a mild natural anti-fungal agent. In Australia, where horsetail is a declared noxious weed, the needles of male casuarina trees are used instead and have been found almost as effective.

Lily Kolisko and her husband, Dr Eugen Kolisko, conducted a series of experiments in the 1920s[2] to verify that these preparations were actually effective and also to verify that the humus material characteristic of each soil preparation could not be formed by any alternative method. For instance, in the case of 500, cow manure was buried in cow horns, earthenware vessels and wooden boxes side by side; only the material in the cow horns converted into a dark, microbially rich, colloidal humus substance. Dr Ehrenfried

Pfeiffer verified their findings and also refined the method of producing the best-quality preparations and the optimum application method and rate in each case.

Some of the preparations are made by processes involving encasing herbs in animal organs. This has caused criticism from some quarters: some people reject the entire Biodynamic method because of this seemingly unusual technique, ignoring the enormously positive results obtained by the extensive use of these preparations worldwide. The fact is that the preparations work with unique effectiveness to transform soils and plants, and scientific studies on Biodynamic farms in Australia and Europe confirm this (see Chapter 4). Many important discoveries over the centuries have been applied to benefit humanity, often for centuries or even millennia before they were understood fully or even partly. A full understanding of (for instance) why cow manure is transformed into such a powerful soil builder only when buried in a cow horn may take some time, but, as with many aspects of nature, scientific progress will no doubt eventually provide the answers.

The following is a brief outline of the method of making the Biodynamic preparations (often

*Figure 2.1. Alex Podolinsky retrieving cow horns.
Photo: Oliver Strewe.*

referred to as 'preps'). Preparation-making is a highly skilled operation, requiring years of training and experience and the best-quality materials. Biodynamic growers obtain the preparations from skilled prep-makers (they are not expensive), just as most musicians buy a good-quality instrument, rather than trying to make their own. The important thing is to learn the correct method of using them and to practise and perfect one's skills in this regard, just as musicians practise their craft. Biodynamics is a very effective method of agriculture, but its effectiveness relies on the use of high-quality Biodynamic preparations. It is a waste of time and effort to use anything less. Those wishing to become involved in prep-making should first become proficient in the Biodynamic method, and then contact their local or national Biodynamic association.

Preparation 500

Cow horns are filled with fresh Biodynamic cow manure (the quality of manure is critical), buried in fertile soil in autumn and dug up in spring. Preparation 500 is made in areas that experience a cold winter. Alex Podolinsky advised that the transformation from raw manure to high-quality 500 occurs only when it is so cold as to preclude microbial activity, and yet the result is a substance containing a very high level of beneficial microbes. He suggested that '500' made in a warm winter area undergoes more of a compost-like transformation than the genuine 500 transformation. A certain number of horns is required for the best 500. A few horns buried by themselves will not develop the 'working together as a whole' as will a larger number (preferably at least 100), and will

Figure 2.2. Alex Podolinsky emptying the 500 out by tapping horns on the metal bar. Photo: Oliver Strewe.

not make very good 500. The two main Bio-dynamic Agricultural Association of Australia (BDAAA) 500 pits hold over 50,000 horns while the other pits each hold tens of thousands of horns. Preparations 502–507 are added to 500 a few months after it has been retrieved from the soil (then left to digest for a further few months before use), to make 'prepared 500', which is considerably more effective than 500 alone ('straight 500') and is the only form of 500 used in most situations on Australian Demeter Biodynamic properties worldwide.

Preparation 501

Good-quality, *clear* quartz crystals are broken down into small pieces and then progressively ground to talcum powder fineness. This is best done between sheets of glass, not by grinding in water. Cow horns filled with a paste made from this fine powder are buried in the soil in spring and retrieved in autumn. A dry, pure white powder is retrieved from the horns when they are dug up.

The Compost Preparations

Preparations 502–507 have a beneficial influence on the composting process, helping regulate temperature and conserving and in some cases increasing elements such as nitrogen. Rudolf Steiner suggested that each preparation has a specific influence on the activity of a particular element or elements, but scientific validation of these relationships is as yet inadequate.

Preparation 502

Dried yarrow flowers (no stalk) are placed in an inflated stag's bladder and exposed to the sun over spring and summer by hanging it in a tree (protected from wildlife by wire cages), then buried in the autumn and retrieved in spring. Preparation 502 has been linked to the activity of sulphur.

Preparation 503

German chamomile flowers, picked at the right stage (neither too young nor too old) and with no stalk, are stuffed into sausage lengths (tied at

Figure 2.3. Picking chamomile flowers (coordinated by Birthe Holt, Australian Demeter Biodynamic method, Herthe community, Denmark). Photo: Birthe Holt.

Figure 2.4. Picking dandelion flowers, Herthe.
Photo: Birthe Holt.

each end) of the small intestine of a cow, buried in autumn and retrieved in spring. Preparation 503 has been linked to the activity of calcium.

Preparation 504

Dried stinging nettle leaves (few stalks) are moistened and packed tightly in earthenware pipes, buried in the soil in autumn and retrieved the following autumn. They are in the soil for a full 12 months. Preparation 504 has been linked to the activity of nitrogen and iron.

Preparation 505

Sheep skulls (or skulls of other farm animals) are filled with finely ground oak bark (*Quercus robur*); they are placed in autumn in a muddy, watery environment through which water flows slowly, and retrieved in spring. Preparation 505 has been linked to the activity of calcium.

Preparation 506

Dandelion flowers – that have not been pollinated by an insect and that are collected early on the morning they first open – are wrapped in the mesentery of a cow, buried in the soil in autumn and retrieved in spring. Preparation 506 has been linked to the activity of silica and potassium.

Preparation 507

Here I shall give more detail[3] because 507 can be very valuable used by itself for its unique warming, healing properties and is not too difficult for individual farmers to make. Its preparation is time consuming, however, since it takes 7 or 8 hours to pick enough valerian flowers to make one wine bottle of 507. Smaller amounts will normally suffice for a single farm.

Valerian flowers are picked when the plants

Figure 2.5. Picked dandelion flowers in baskets, Herthe. Photo: Birthe Holt.

are in full bloom. Only newly opened flowers are picked, with no green parts (all petal and no sepal). Work over individual clusters each day to ensure that only newly opened flowers are picked. The flowers are infused in rainwater in glass jars (generally in the proportion two-thirds flowers to one-third water, but in years when the flowers are particularly strongly scented, use a bit more water). Add some flowers to the jar, then some water, then stir the flowers in.

Figure 2.6. Drying dandelion flowers, Herthe. Photo: Birthe Holt.

Figure 2.7. Dandelion flowers, wrapped in mesentery parcels, ready for burying, Herthe. The mesentery is translucent, an important factor for high-quality 506. Photo: Birthe Holt.

Add more flowers and more water and stir again. Repeat until the level is 10mm below the top of the jar. Don't stir or shake the jar any more. Put a metal lid tightly on the jar. Tie a string around the neck of each jar and hang in a tree that allows gentle sunshine through. Loosen the lids a little once they are securely hung. In countries where the sun is less intense, more light can be allowed to contact the jars. The jars are left for a few days or until the liquid develops a refined scent and a light honey colour. The liquid is then strained through a coarse strainer and then through a circular filter paper (Whatman No. 1) – that is folded in half and in half and in half again before it is opened out – in a funnel in a glass bottle (it takes some time to filter). Fill the bottles completely, leaving no room for any air, put in a stopper and seal with hot beeswax. Store the bottles on their sides in a cool dark place until needed. Quantities smaller than that of a wine bottle can be stored in one section of the compost preparation storage box (see p. 99). Preparation 507 has been linked to the activity of phosphorus. It has also been suggested that 507 has a unique warming influence; it is used on crops threatened by severe, damaging cold and after severe storm damage. More research is needed to fully validate this.

Once made, the preparations must be stored carefully until needed, so that they retain their effectiveness. Preparations 500 and 502–506 are stored in peat-moss-insulated boxes (see p. 68). Preparation 501 is stored in a jar on a windowsill exposed to morning sun (turned occasionally), while, as noted, bottles of 507 are stored on their sides in a cool, dark place or in the compost preparation storage box.

High-quality Biodynamic preparations must be moist and colloidal, except for 501, 505 and 507. Preparation 501 (quartz) is a dry powder, and 507 (valerian) a liquid. Preparation 505 (oak bark) doesn't have the same colloidal nature as the other solid preps; it must still be kept moist at all times, requiring more water than the others (add a bit of water weekly) and needing to be inverted every 3 to 7 days (jar lid is temporarily tightened while one does this) so that the water that has gravitated to the bottom of the prep, making the bottom too wet and leaving the top too dry, will redistribute itself evenly. Preparation 506 (dandelion) needs special care for the first few months after lifting, having a tendency to settle down and pack together. Left to itself, it will become a non-colloidal anaerobic 'swamp'. For the first few months after lifting it should be turned over or shaken in the jar (only half to three-quarters full) every day. After a few months it settles down into a stable colloidal state and doesn't need further turning or shaking.[4] All the solid preps need ongoing monitoring; pure water is added as needed to preserve the same moist colloidiness as when they were first lifted. Dried-out preparations are ineffective.

FOOTNOTES FOR CHAPTER 2

1. Referred to as 'Biodynamic preparation 508' in some countries.
2. Their results were presented in Lily Kolisko and Eugen Kolisko, *Agriculture of Tomorrow*, Kolisko Archive, Stroud, 1978.
3. As taught by Alex Podolinsky and Frances Porter.
4. These points are as taught by Alex Podolinsky.

Chapter 3

A Brief History of Agriculture

Before going further into the Biodynamic method, let's pause to consider its place in the evolution of agriculture. Whichever agricultural method we use, it is useful to understand its place in world agricultural development, and its consequences.

The dawn of agriculture occurred only relatively recently in human history. Hunter-gatherer societies began domesticating cattle, sheep, goats and other animals and slowly transitioned into nomadic herding lifestyles. Around 11,000 to 12,000 years ago, in the Fertile Crescent (from modern-day Israel to the head of the Persian Gulf), some groups began to settle permanently and domesticate and cultivate wild plants. The first crops domesticated and improved included einkorn and emmer wheat, barley, legumes, grapes, melons, dates and almonds. We must all be immensely grateful for the vision and skill of these agricultural pioneers.

The Tigris, Euphrates and Nile rivers annually deposited nutrient-rich silt over their flood plains, providing wonderful fertility for the newly developing settled agriculture. In other areas of the world, simple 'slash and burn' agriculture developed: patches of bush were burnt and cleared and crops grown. When the soil became exhausted, the area would be left to regrow and recover and a new patch would be cleared.

As populations grew and new areas had to be developed for agriculture, more sophistication in farming was required. Annual silt deposition only occurred in limited areas, and not enough land was available for slash and burn. Gradually, our ancestors developed increasingly sophisticated methods of replenishing soil humus by such techniques as fallowing, crop rotation, and the conservation and reuse of all crop, animal and human wastes.

Whether in Europe, Asia, South America or elsewhere, organic agriculture developed to the point where soils could be replenished with humus year after year and crops could continue to be successfully grown, supporting the development of great civilisations. Meticulous record-keeping in European monasteries tells a story of high productivity and relative freedom from pests and diseases.[1]

From the 18th century onwards, scientists focused more and more on understanding how plants feed and how soils should be managed. In the early 19th century, a German agronomist,

Albrecht Thaer, who established the Royal Prussian Academy of Agriculture, carried out extensive and meticulous experiments and field trials in all aspects of agriculture. His most important conclusion was that soil fertility depended on the level of soil humus and that humus constituted plant food.[2] He regarded humus as a biological-functional performance of the earth that should be viewed holistically. He also believed that inorganic salts were unnecessary to plant nutrition.

The Pivotal Point in Agriculture

The major departure from traditional organic methods came in the early 1840s: German chemist, Justus von Liebig, discovered that plant roots can only absorb elements in solution. This important discovery quickly led to the commercial production and promotion of water-soluble 'artificial fertilisers' – superphosphate in the 1840s and nitrogenous fertilisers in the 1890s.

Late in his life, von Liebig realised that water solubility was only part of the picture and that humus was, in fact, extremely important in plant feeding. However, by then commercial forces were in full flight, and there was money to be made from artificial fertilisers. Von Liebig's discovery led to most scientists discarding Albrecht Thaer's humus theory, completely ignoring his voluminous research results.

In the latter half of the 19th century, farmers using water-soluble fertilisers certainly experienced higher crop yields. At this stage, nitrogen, phosphorus and potassium – NPK – were the only water-soluble fertilisers used. But, gradually, many noticed that the food they were producing was not as flavourful as it had

been before, that pest and disease problems were increasing and that seed vitality and animal fertility were declining. Many farmers and farm estate owners, concerned by these developments, began to search for solutions.

Biodynamics

In the early 1920s, a group of farm estate owners approached Dr Rudolf Steiner, an Austrian philosopher and theosophist who had established a reputation for finding practical solutions in many spheres of human endeavour, and asked him to help them with their agricultural problems. In 1924 he found time to give a series of eight lectures on agriculture, introducing many new ideas for rejuvenating soils, plants and animals, using natural 'organic' methods, including the making and use of a series of preparations that would greatly improve soil fertility as well as assisting light metabolism in plants. Among his many suggestions was that plants' growth requires not just NPK but also a wide variety of other elements in very small amounts (trace elements); this was at least a decade before other scientists began to realise the latter's importance. Steiner insisted on a thorough scientific evaluation of all his ideas. Scientists such as Dr Ehrenfried Pfeiffer, Dr Eugen Kolisko and Lily Kolisko carried out the research validating Steiner's key suggestions. The method developed rather slowly in Europe, before Hitler's rise to power stopped all development.

Australian Biodynamics

In 1952, Alex Podolinsky, who had emigrated to Australia after the Second World War, started

Biodynamic farming on a poor, shaly property at Wonga Park (now a suburb of Melbourne). His aim was to develop the Biodynamic method of agriculture to suit modern, broadacre, low-labour farming conditions. He ran a highly productive Biodynamic cherry orchard and also ran some cattle. In bad years for rot, his cherries were sometimes the only ones on the market; spores from neighbouring conventional orchards blew through, but his cherries were unaffected.

Podolinsky developed many of the fundamentals of Biodynamic agriculture at Wonga Park, such as deep ripping to loosen compacted subsoils, and the use of chisel ploughs under stress to 'hammer' and loosen subsoils. Right from the start, rotational grazing through many paddocks, and strip grazing[3] were used. Podolinsky began making Biodynamic preparations following Pfeiffer's methods and improved on Pfeiffer's technique with several of them, notably 504 (stinging nettle). This was acknowledged by Pfeiffer, with whom Podolinsky exchanged preps for comparison from time to time.

In the mid-1950s, Podolinsky moved to a degraded potato farm at Powelltown, Victoria, which he converted for dairying. Within a few years he had totally rejuvenated it without any fertiliser inputs and had increased the organic matter content in the top 100mm from 0.9% to 11.4%. He began topping Victorian dairy production figures, and many farmers and Agriculture Department officials became interested. He began to train other farmers and in the early 1960s founded the BDAAA in Victoria with 27 of these farmers. Podolinsky developed the Biodynamic method together with a growing cohort of highly creative and resourceful farmers to the extent that by the early 1980s it had been adopted by Australian farmers on hundreds of thousands of acres. In 1982, a national TV feature on Podolinsky and Biodynamics was screened. As a result, Podolinsky received over 6,000 letters from farmers, and many more converted to Biodynamics.

The many key developments that underpin the Australian Demeter Biodynamic method include:

- The key understanding that plants should be fed 'within nature's organisation' (see pp. 4–5).
- Podolinsky's development of a series of lectures on Biodynamic agriculture which he presented all over Australia and worldwide. These lectures presented Biodynamics as developed and practised in Australia. They spoke directly to farmers' experience and were instrumental in the wide adoption of

Figure 3.1. Rudolf Steiner.

Figure 3.2. Alex Podolinsky. Reproduced from the Podolinsky Collection with kind permission.

Biodynamic farming in Australia. The lectures were published as *Biodynamic Agriculture, Introductory Lectures Volumes 1, 2 and 3* and have been translated into many other languages (including German, French, Italian, Russian and Chinese).

- A further five books on Biodynamics, based on lectures by Podolinsky, provided further developments and insights.[4]

- Adherence to a strictly science-based approach in the development of the method. This does not preclude inspirational insights (from Alex or any of the farmers) but requires all such insights to be thoroughly tested. It entails absolutely meticulous preparation-making techniques (stemming originally from Pfeiffer's research and recommendations, with some uniquely Australian improvements), careful storage and handling of the preparations, using the exact amounts determined by research findings, and careful attention to application under correct environmental conditions to maximise the preparations' effectiveness.

- The development of 'prepared 500', which incorporates the six Biodynamic compost preparations and which, after extensive testing, proved significantly more effective than 500 alone.

- The development of a stirring machine (to oxygenate and activate 500 before spraying) that replicates the quality of hand stirring. This enables 500 to be applied quickly and efficiently over large-acreage farms and has been a major factor in the wide uptake of Biodynamics in Australia and, increasingly, worldwide. It was invented by Biodynamic farmer Kevin Twigg, together with Murray Gartner (who did the electrical work). Later, Biodynamic farmer Trevor Hatch invented a hydraulically driven version. In his latter years, Alex Podolinsky called for a further refinement: a stirring machine whose stirring arms started stirring in the middle, but spread outwards as the vortex developed. Melbourne Biodynamic grower, Rod Turner has built a prototype that does exactly that.

- The development of spray rigs capable of spraying 500 over large acreages quickly.

- The development of caring cultivation techniques and implements that would complement the soil structure development resulting from the use of 500. This includes Podolinsky's world-first use, in the early 1950s, of deep ripping (and also the 'chiselling' action of a chisel plough, the rapid back-and-forth vibration of each tine as it moves through the soil) to relieve soil compaction, and the development of the (Italian) Agrilatina and (Australian) Rehabilitator ploughs, based on Podolinsky's visualisation and description of the ideal plough.

- The development of an 'active viewing' technique that enables farmers to actively engage with the life of their farm environment, crops and animals and so gain valuable insights. This is at once objective and subjective, requiring the active (subjective) efforts of the individual while using the objective method.[5]

Modern Organics

Modern organic agriculture started well after Biodynamics, in 1940, with the publication of two key books, Albert Howard's *An Agricultural Testament* and Lord James Northbourne's *Look to the Land*. Howard's book, based on his work as a director of agricultural research centres in India from 1905 to 1931, stressed the importance of returning all crop wastes to the soil to build soil fertility and humus, and described detailed methods of compost-making. Northbourne coined the word 'organic' in relation to organic agriculture; he referred to the farm as an 'organic whole'. Many of Northbourne's ideas stemmed from Steiner's agricultural lectures.

The fundamental limitation of ordinary organic agriculture is the absence of the most powerful soil fertility developers, the Biodynamic preparations. In the absence of their influence, soil development takes much longer, and large amounts of organic fertilisers are often imported on to farms.

Conventional Agriculture: The Future

I would contend that conventional agriculture has no long-term future. It has, for 180 years, been pursuing an agricultural theory that seems deeply flawed. It damages soil fertility and structure, feeds plants indiscriminately, outside nature's organisation, and consequently requires spraying toxic chemicals to keep plants alive. The safety testing of these chemicals is lacking in scientific rigour and is mostly controlled by the very companies that produce the chemicals and have a vested interest in their being declared safe. The use of such chemicals causes damage to soils, to the environment, to the health of the farmers using them and to that of consumers. It is, in my opinion, a dead end in the development of agriculture.

FOOTNOTES FOR CHAPTER 3

1. Some imports of manure, hay etc. from local farms were used by monasteries.

2. Sixty years ago, Alex Podolinsky demonstrated that colloidal humus can be completely absorbed by plant roots. He buried a glass jar filled with colloidal humus 100mm deep in active, 'biologically working pasture' in spring. After 6 weeks, the jar was completely filled with white hair feeder roots and all humus had completely vanished (*Biodynamic Agriculture Introductory Lectures Volume 1*, Gavemer, Sydney, 1985, p. 23).

3. Dividing fields into smaller areas with electric wires, and grazing strips progressively, to more efficiently manage the grass.

4. *Bio-Dynamics Agriculture of the Future* (2000), *Living Agriculture* (2000), *Living Knowledge* (2002), *Ad Humanitatem* (2003), *FiBl Lecture* (2004) – all published by BDAAA, Powelltown, Victoria.

5. See Alex Podolinsky, *Active Perception*, Gavemer, Sydney, 1990.

Chapter 4

The Scientific Basis of Biodynamics

From the very start, Biodynamic development has been characterised by systematic scientific evaluation. There is no doubt that Rudolf Steiner's suggestions about the making of the soil fertility and light metabolism preparations seem strange. However, this is entirely irrelevant to their objective evaluation. The unusual suggestions, such as putting cow manure in cow horns and burying them over winter to make 500, together with Steiner's theosophical teachings and esoteric world view, have led some scientists to ridicule the Biodynamic method, completely ignoring the science behind it. Let's look at the facts and leave the hysteria to one side.

The scientific method starts with a question that you want to find answers to. After conducting observations of the phenomena involved and reviewing any related studies, a hypothesis (an educated guess) is formed that may potentially explain the phenomena in question. Experiments are then conducted to establish the validity or otherwise of your hypothesis. The results are analysed and conclusions drawn based on the results – a rational, orderly process. However, as eloquently put by Fred Hoyle[1] and the astronomer Raymond Arthur Littleton, the process is often

ignited by 'intuitive insight or other inexplicable inspiration'.[2] Throughout the history of science, many major discoveries have come as a result of some sudden flash of inspiration that has led to the development of a hypothesis and subsequent testing of the hypothesis. The source of the inspiration is completely irrelevant, as is the belief system of the person advancing the hypothesis. Whether it be Steiner's inspirational agricultural suggestions, August Kekulé's dreams and visions leading to the discovery of the molecular structure of organic chemicals, or Alex Podolinsky's vision of plant functioning, it is the subsequent testing that establishes the validity or otherwise of the hypotheses in question.

Many of the most influential scientists of the last few centuries were religious or believed in some higher power, including Charles Darwin (biology/evolution), Marie Curie (chemistry), Isaac Newton (physics), Gregor Mendel (genetics) and Srinivasa Ramanujan (mathematics). According to a 2009 survey, 51% of American scientists believe in God or some higher power. Just as there is no justification for ridiculing these scientists' ground-breaking discoveries because of their personal belief systems, there is no justification for ridiculing

Steiner's agricultural theories because of his theosophical writings.

Steiner – whose doctoral dissertation was entitled 'Truth and Science' – insisted on thorough scientific testing of his agricultural suggestions. During his tertiary education, he carried out scientific testing of Johann Wolfgang von Goethe's theory of colour, confirming many of Goethe's findings. His extensive study and knowledge of Goethe's scientific works led to his appointment as editor of these works for publication in the German National Literature Series. Before the 1924 agriculture lectures commenced, Steiner had requested that Ehrenfried Pfeiffer and others undertake an initial trial of burying cow manure in cow horns over the autumn/winter period. That trial provided early confirmation of Steiner's suggestion about making 500.

Steiner had met Pfeiffer when the latter was a young university student, and encouraged him to concentrate on studying the sciences,

particularly biology, chemistry and physics. Pfeiffer was entrusted with the task of evaluating Steiner's agricultural suggestions, particularly those relating to the making of the preparations. He confirmed that the preparations were effective and refined the method of making them as well as the most effective application methods and quantities.

Steiner also asked Lily Kolisko (a laboratory technician) and her husband, Dr Eugen Kolisko (a medical doctor and lecturer in medical chemistry), to similarly test the preparations. They too validated Steiner's suggestions and went on to study other areas such as the influence of the moon and planets and the effect of very dilute substances on plant growth, among other topics. Lily continued her studies for over 30 years.

In Australia, Podolinsky founded the Bio-Dynamic Research Institute (BDRI) in 1957 with a colleague, Andrew Sargood. Together they carried out many tests on soils and on Biodynamic developments. Many field trials

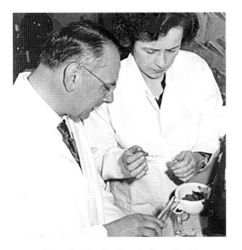

Figure 4.1. Ehrenfried Pfeiffer (left). All efforts were made to find the copyright holder; the publisher would welcome further information.

Figure 4.2. Lily Kolisko. Reproduced by kind permission of Andrew Clunies-Ross, Kolisko's grandson. www.koliskoarchive.com

were conducted. The newly invented stirring machine was thoroughly evaluated and found to produce better soil development results than hand stirring for any quantity of water over 12 imperial gallons (54 litres).[3] Podolinsky's new development, prepared 500, was found to be considerably more effective than straight 500 and became the only form of 500 used in most situations in Australia (and worldwide on farms following the Australian Demeter Biodynamic method). Podolinsky followed Pfeiffer's preparation-making recommendations, improving on some, as acknowledged by Pfeiffer, with whom Podolinsky exchanged preparation samples periodically until Pfeiffer's death in 1961.

Tests carried out by the BDRI showed that, using no fertiliser inputs at all, Podolinsky increased the soil organic matter level at Powelltown, in his first six years there, from 0.9% to 11.4% in the top 100mm and from 0% to 2.4% at 1000mm. Victorian Agriculture Department scientists reported that this had resulted in the locking up of 1,614 tonnes of carbon dioxide per hectare over the 6-year period.[4]

Independent research has added to the validation of the method in Australia. J.A. Lytton-Hitchings et al. studied matched pairs of Victorian Biodynamic and conventional dairy farms. They found that the Biodynamic farms had greater soil macro-porosity, better structured soil, greater air-filled porosity and greater organic matter content. They concluded that 'these more favourable soil properties of the bio-dynamic soil have the potential to allow faster infiltration, less surface runoff,

less waterlogging, deeper soil exploration by plant roots and a longer interval between irrigations'.[5] Parker[6] and Cock,[7] too, found better soil structure on Biodynamic farms.

In 1991, a study of ten matched pairs of Biodynamic and conventional dairy farms was carried out by the Victorian Agriculture Department, overseen by soil researcher Doug Small and veterinarian Dr John McDonald. The results were reported at an Australian Institute of Agricultural Science conference in 1993. In summary, the Biodynamic farms (which were using no fertiliser at all) were found to have better soil structure, similar pasture composition, slightly lower soil phosphorus and soil nitrate nitrogen levels, and other minerals at similar levels; Biodynamic cows didn't suffer from bloat, and the incidence of metabolic disorders was lower; irrigation was much less frequent on Biodynamic farms; lower levels of selenium were found in conventional cows, lower levels of phosphorus in some Biodynamic cows; Biodynamic cows were more fertile; Biodynamic cows remained productive longer; grain feeding was 600% higher on conventional farms; milk production was higher on conventional farms, but costs were lower on Biodynamic farms. Net returns were somewhat higher on conventional farms, but if off-farm environmental effects were taken into account (nutrient run-off has been found to be up to 30 times higher on conventional farms, causing blue-green algae and other environmental problems), it was suggested, the total economic benefit would likely be in favour of Biodynamics. McDonald stressed to me just how important was the higher level of selenium in the Biodynamic cows, relating this to their higher fertility levels and also suggesting that it may have significant consequences for human fertility.

A study comparing Biodynamic with conventional soils was carried out by Eric Frescher as part of his Bachelor of Civil Engineering at LaTrobe University.[8] The Biodynamic farm studied had previously demonstrated considerable reduction in salt-affected areas over a 5-year period. During the same period, the neighbouring conventional farm studied had demonstrated significant expansion of salt-affected areas. Frescher's results showed that the Biodynamic soil was less compacted and more porous than the conventional soil in both a dry and a wet state; water infiltration on the Biodynamic soil was four times higher for the first ten minutes and two and a half times higher after that; the Biodynamic soil was capable of holding twice the amount of soil moisture and contained twice the amount of carbon; the Biodynamic soil contained numerous aggregates and was softer, whereas the conventional soil contained very few aggregates and was harder; the Biodynamic soil was five times less acid than the conventional soil; the Biodynamic soil contained six times more nitrate nitrogen in the top 50mm and at 200mm contained 10ppm nitrate nitrogen compared with zero on the conventional farm; and, in dry conditions, the soils showed similar levels of biological activity (as measured by microorganism respiration

Table 4.1.

Soil characteristic examined	Conventional (%)	Organic (%)	Biodynamic (%)
Microorganism activity [a]	100	143	161
Weight of earthworms [b]	100	130	140
Root penetration intensity [c]	100	103	157
Soil crumb stability [d]	100	104	125
Nitrous oxide emissions [e]	100	75	57

a. J. Raupp, 'Wirkungen der biologisch-dynamischen Präparate im Langzeit-Düngungsversuch', *Lebendige Erde*, 5 (2001), pp. 42–44.

b. L. Pfiffner and P. Mäder, 'Effects of Biodynamic, Organic and Conventional Production Systems in Earthworm Populations', *Biological Agriculture & Horticulture*, 15 (1997), pp. 3–10.

c. J. Bachinger, 'Der Einfluss unterschiedlicher Düngungsarten (mineralisch, organisch, biologisch-dynamisch) auf die zeitliche Dynamik und räumliche Verteilung von bodenchemischen und mikrobiologischen Parametern der C- und N-Dynamik sowie auf das Pflanzen- und Wurzelwachstum von Winteroggen', *Schriftenreihe Institut für biologisch-dynamische Forschung*, 7 (1996).

d. S. Siegrist, D. Schaub, L. Pfiffner and P. Mäder, 'Does Organic Agriculture Reduce Soil Erodibility? The Results of a Long-Term Field Study on Loess in Switzerland', *Agriculture, Ecosystems & Environment*, 69 (1998), pp. 253–264.

e. E.M. Minarsch, A. Gattinger, C. Skinner and P. Mäder, 'Potenziale des Ökolandbaus in Zeiten des Kimawandels: Geringere Lachgas Emissionen vor allem auf biodynamischen Ackerböden', *Lebendige Erde*, 2 (2020), pp. 40–45.

rates), but in wet conditions the Biodynamic soil biological activity was far higher.

European Studies

The Research Institute of Organic Agriculture (Forschungsinstitut für Biologischen Landbau [FiBL], Switzerland) ran trials comparing Biodynamic, organic and conventional growing methods and published between 1996 and 2020. A selection of results is shown in Table 4.1, expressed as a comparison with conventional growing methods (100%).

Why Do Biodynamic Growers Need Fewer Fertiliser Inputs?

The ideal of many organic and Biodynamic farmers has been to achieve a closed system, that is, to bring in no fertiliser inputs at all from outside. This is partly based on the fact that European organic farms of the past operated successfully for centuries as almost closed systems. However, pre-mechanisation, these farms had large numbers of people living on them (farm workers and their families). Most of the farm production went to feed these families, relatively little being exported. All farm wastes, including animal and human manure, were recycled by composting and was spread on melting snow in spring to maintain soil humus levels and fertility. In the early period of Demeter Biodynamic development in Australia, many farmers aspired to operate their farms as closed systems and applied no fertiliser for decades. They were able to maintain good production levels (Alex Podolinsky brought in no fertiliser for 40 years, until prolonged drought necessitated an application of reactive

rock phosphate), but this regime could not be maintained for ever. It was eventually realised that modern organic farming is quite different from historical organic farming. Now, usually, only one family lives on and operates a large farm, sometimes running into thousands of acres, and must export large quantities of produce to survive economically. Some fertiliser inputs may well be required to complement the work of the soil biology activator preparations and general Biodynamic management. Nevertheless, experience shows that Biodynamic farmers require far less fertiliser inputs than conventional farmers, and considerably less than most organic farmers. Why is this?

Why Biodynamics Works

The reason that Biodynamics works so well is that every aspect of the method, based on scientific understanding, meticulous experiments and extensive field trials, is focused on building and protecting the essential soil life, whose collective function in nature is to convert organic matter and any free soluble elements into stable colloidal humus. This feeds plants as nature intended, under the jurisdiction of the sun, not indiscriminately through the soil water.

Myriad living organisms and associated 'biological catalysts' participate in this soil life:

- *Soil animals*, including micro-fauna such as protozoa; meso-fauna such as mites and springtails; and macro-fauna such as ants, earthworms, beetles and termites.
- *Soil fungi*, active in the early stages of breakdown of organic matter.
- *Mycorrhizal fungi*, a specialised group

of fungi that form mutually beneficial associations with plant roots.

- *Soil bacteria*, single-celled organisms that can reproduce and proliferate very quickly. They include: *Actinobacteria* (active in converting organic matter into humus); independent nitrogen-fixing bacteria, which convert ammonium into nitrates that are incorporated into the soil humus; *Azobacter*, free-living bacteria that convert atmospheric nitrogen into forms useable by plants; rhizobia, bacteria that form symbiotic relationships with the roots of leguminous plants and take nitrogen from the air and convert it into a form useable by plants.
- *Biological catalysts such as enzymes* and co-enzymes, which are synthesised by plants, animals, fungi and bacteria. They play a key role in the breakdown of organic material and its conversion into colloidal humus, the cycling of nutrients and the building of soil structure.

The Biodynamic preparations are the most powerful (and unique worldwide) stimulators of this overall biological activity in the soil, and, when combined with associated biological agricultural practices, enable farmers and gardeners to produce food of the highest quality with absolutely minimal inputs.

Biological Transmutations

On the basis of an expanding body of research, it is becoming clear (though not yet accepted by many scientists) that, in a living biological system, elements can transmute into other elements. This is the inescapable conclusion of much meticulous research and many observations that are inconsistent with the generally accepted view that an element is a substance that cannot be broken down by chemical means. Rudolf Steiner suggested in 1924 that transmutation of elements was occurring in nature.

Research that supports this idea includes the following: in 1799, Louis Vauquelin found that a hen excreted 5 times more lime than it ingested; in 1822, Prout discovered that lime in an incubating chicken egg increases overall; in 1831, Choubard found that germinated seeds contained minerals that were not originally present in the seeds; in 1844, Vogel found that, after germination, watercress contained more sulphur than was in the seeds. These results strongly suggest that under certain conditions, elements are able to transmute into other elements.

In 1879, Albrecht von Herzeele published the results of many experiments[9] that strongly supported the idea that elements can change from one into another. Some of the reactions he demonstrated included:

$$H_2CO_3 \rightarrow Mg \rightarrow Ca \rightarrow P \rightarrow S \rightarrow N \rightarrow K$$

How is this possible? Conventional understanding is that elements can combine in chemical reactions to form molecules, and that molecules can break down into individual elements, but that elements themselves cannot normally change.

A chemical element is a pure substance composed of one type of atom. Elements are distinguished from one another by the number of protons (positively charged particles) in the atomic nucleus. The lightest element, hydrogen, has one proton in its nucleus, giving it an atomic number of 1. Uranium is the heaviest naturally occurring element, with 92 protons and therefore the atomic number 92. Biological transmutation occurs when an element merges with another element, combining their protons in a single new nucleus, so that the two become a third element, or, conversely, when an element breaks down into two new elements with separate nuclei.

It is accepted by scientists that in certain circumstances elements can change: when a cosmic ray neutron (electrically neutral particle) hits a nitrogen atom (seven protons), for example, it breaks down into carbon (six protons) and hydrogen (one proton). Isotopes (variations in the number of neutrons in an atomic nucleus) are involved but we won't go into that here.

The element uranium breaks down (very slowly) in nature in the following sequence (again involving isotopes): uranium to thorium to protactinium to uranium (a different isotope) to radium to radon to polonium to lead to bismuth, and, after fluctuating between various isotopes of lead, bismuth and polonium, eventually decays into a stable form of lead.

Professor Louis Kervran (University of Paris) began publishing the results of his experiments on what he called 'biological transmutations' in 1959. He found that the most important and abundant biological transmutations occur among the first 20 elements of the periodic table and to a lesser extent with the next ten. Many transmutations

were found to occur. Although the exact mechanism of biological transmutation is still to be discovered, Kervran suggested that enzymes are probably integral to the process. Scientists who have confirmed some of his findings include: Hisatoki Komaki, Pierre Baranger, J.E. Zündel, Panos T. Pappas, Jean-Paul Biberian, Vladimir Vysotskii, Edwin Engel and Rudolf Gruber.

Leaving aside the issue of different isotopes, and noting the atomic number only, some of the biological transmutations that have been discovered (many of which are reversible) include:[10]

$$_{11}Na + {}_8O = {}_{19}K \qquad _{19}K + {}_1H = {}_{20}Ca$$
$$_{12}Mg + {}_8O = {}_{20}Ca \qquad _{14}Si + {}_6C = {}_{20}Ca$$
$$_{11}Na + {}_1H = {}_{12}Mg \qquad _{11}Na = {}_3Li + {}_8O$$
$$_{12}Mg + {}_3Li = {}_{15}P \qquad _{17}Cl - {}_8O = {}_9F$$
$$_{15}P = {}_6C + {}_9F \qquad _8O + {}_8O = {}_{16}S$$
$$_{26}Fe - {}_1H = {}_{25}Mn \qquad _7N + {}_{12}Mg = {}_{19}K$$

These are only a sample of the many reactions that have been shown to occur.[11] The implications for farming, diet and medicine are intriguing. Instead of trying to replace elements that appear to be lacking we might do better to carefully consider biological transmutations. For instance, it may well be (there is some evidence[12]) that calcium deficiency in humans could be better treated with magnesium or organic silica (both of which can biologically transmute into calcium) than with calcium supplements. In agriculture, potassium can be formed from the transmutation of calcium (in combination with hydrogen), and so on.

The very idea of 'soil analysis' is called into question when we are dealing with the biologically active soils found on Biodynamic and organic farms and gardens. Where soil life is abundant, many active cycles occur. Nature is constantly trying to achieve balance, to redress imbalance, so that healthy plant growth can continue. A soil test is a snapshot in time. One week after the snapshot, the situation may have changed considerably. In 2004, the Australian Soil and Plant Testing Council sent standardised soil samples to 18 laboratories across Australia. The results for nitrogen, phosphorus and potassium varied so dramatically that laboratory standards were called into question, even though the same testing methodology was used. Could it be that, once the samples were divided up, variations in temperature, elevation, pressure, microbial and fungal content, or other factors, could have caused the wide divergence via biological transmutation?

Although soil tests can still be useful and are used from time to time by Biodynamic farmers, it is perhaps more important to first look at the whole living system on a farm to assess the status of the soil, including how well or poorly plants are growing, the predominance of particular grass or weed species, the health status of animals, and so on, than to rely solely on potentially unreliable soil tests.

Kervran makes the point that in agriculture, for biological transmutations to work, the soil must be alive, rich in microorganisms, and that adequate humus must be present. He also comments that all plants are different. For instance, some plants can make their own calcium in a calcium-deficient soil whereas others cannot. The same

is true of other major and minor elements. Each plant has different root types and structures and releases different exudates from its roots. Each plant lives in, fosters and is fostered by a unique community of bacteria and fungi.

We simply don't know how many different species of living things are in the soil or understand the complex interrelationships that exist, or how they vary from plant to plant. We don't know the full range of enzymes and co-enzymes that are produced by plants and soil macro- and micro-life or the biological transmutations they, the microorganisms and possibly creatures such as worms may foster.

The correct use of properly made and stored Biodynamic preparations enormously enhances soil biological activity. Organic matter levels in soils rise significantly. Microbial activity is greatly enhanced, root growth fostered and colloidal humus formation promoted. As a direct result of this enhanced activity, soil colour darkens progressively and soil structure improves. Actively growing roots, in conjunction with soil biology, create and develop soil structure. Soils become deeper, even heavy clays becoming friable soil in time. Such dramatic soil changes, which soil scientists have stated would take nature alone hundreds if not thousands of years to achieve, have only been achieved through the use of the Biodynamic preparations.

The soil spray 500, if properly made, is a very concentrated source of microbes, as are the six compost preparations that are used in compost heaps and added to 500 to make the even more potent prepared 500. Pfeiffer discovered several novel species of microbes in the preparations.

Sprayed on moist soil that is warm (or at least not too cold), the Biodynamic preparations have a unique and dramatic soil-enlivening effect. The conditions for a rich soil ecosystem, which facilitates transmutation of elements, are well and truly established by their correct use.

Sun and Air

In discussing plant growth we tend to focus on soil factors and forget that without sun and air there would be no plants at all. The sun provides the primal energy for photosynthesis, the only truly original manufacturing process on earth. Air provides the carbon dioxide necessary for photosynthesis to occur, as well as the nitrogen that is brought into soils by *Azobacter* and rhizobia in conjunction with plants. We must also remember that plant leaves can take in other nutrients from the air; all elements are present in air, even if in minute quantities. We still don't understand how relevant this is to plant nutrition, although the fact that leaves can take in nutrients has been abused by proponents of foliar feeding, who apply nutrients via foliar sprays, thus subverting the natural, sun-directed plant feeding process.

Biodynamic management combines:

- the harmonious functioning of the soil engendered by the use of the Biodynamic preparations, encompassing unique soil biology development, humus formation and soil structuring;
- the biological transmutation of elements enabled by the highly active soil biology, giving the soil a high degree of adaptability and supporting the healthy growth of plants;

- a comprehensive system of natural farming techniques, including careful cultivation, rotational grazing of pastures, composting, green manuring and much more.

These factors together provide a plausible explanation for the success of and the much lower input requirements of Biodynamic farms. All of these points have been the subject of scientific studies and extensive field trials. There is, however, one further important aspect that should be considered, one whose exploration is in its infancy and deserves far more scientific scrutiny.

The Concept of Formative Forces

From ancient times, peoples at all stages of civilisation have shared a common belief that living things have as their basis a non-physical 'formative force' that organises and holds the physical components of organisms in coherent, living forms. Different cultures have different names for it – for instance, 'prana' (India), 'chi' (China) and 'ki' (Japan). In the West it has been called 'formative force' and 'etheric body', among other names. It has been written about by philosophers including the ancient Greeks and Lao Tzu, the Chinese founder of Taoism. Most traditional and integrative medical disciplines – including Ayurveda, traditional Chinese medicine, naturopathy, chiropractic and homeopathy – have vitalism as their foundation, as do yoga and many martial arts. Many scientists in the 17th, 18th and 19th centuries were vitalists. Even in the 20th and 21st centuries, some scientists have given credence to this theory. Theoretical physicist Max Planck (1858–1947) was one.

The concept of formative forces was strenuously attacked during the 18th and 19th centuries. When Friedrich Wöhler succeeded in synthesising the organic chemical urea in 1828, this was trumpeted as the death knell of the theory of vitalism, despite the fact that urea is not a living organism. Today most but not all biologists believe that physical and chemical principles are sufficient to account for life despite its many, as yet unexplainable, mysteries.

In his 1924 agricultural lecture series, Steiner suggested that forces, including formative forces, would be concentrated in the Biodynamic preparations and that, when the preparations were applied, these forces would bring a unique enlivening influence to soil and plants as well as to their microbial content. Among the many suggestions he made to the scientists he appointed to investigate his agricultural theory was that the presence and quality of formative forces could be tested by the use of chromatography in connection with capillary analysis. The technique was called 'capillary dynamolysis'. A substance to be tested is liquefied and allowed to rise in a filter paper. The filter paper is then dried and a light-sensitive chemical such as silver nitrate or gold chloride is allowed to be drawn up by the same filter paper. On the paper's exposure to light, a picture emerges that Steiner maintained would give indications as to the presence and quality of formative force in the test subject. Later versions of the method include 'sensitive crystallisation', in which a liquid extract is mixed with copper chloride, poured into a glass dish in a temperature- and humidity-controlled, vibration-proof cabinet

Figure 4.3. Fresh rhubarb juice followed by 1% silver nitrate. Reproduced from Lily Kolisko and Eugen Kolisko, Agriculture of Tomorrow, *Kolisko Archive, Stroud, 1978, by kind permission of Andrew Clunies-Ross. www.koliskoarchive.com*

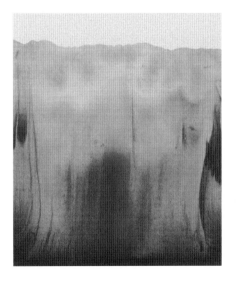

Figure 4.4. Preserved rhubarb juice followed by 1% silver nitrate. Reproduced from Lily Kolisko and Eugen Kolisko, Agriculture of Tomorrow, *Kolisko Archive, Stroud, 1978, by kind permission of Andrew Clunies-Ross. www.koliskoarchive.com*

and left to crystallise. Although these methods may seem tenuous to some scientists, they are not dissimilar to current efforts deep in an Australian gold mine to detect very ephemeral traces of the theoretical, invisible and hitherto unmeasurable 'dark matter' that is purported to constitute 30.1% of the universe. The rest is, in theory, made up of 69.4% invisible and unmeasurable 'dark energy' and 0.5% actual visible matter.

It has been suggested that the Biodynamic soil/compost preparations not only work at a physical level, fostering microbial activity, humus formation and root growth, but also harmonise the working of 'life forces' in soil and plants, facilitating the optimum development of soil biology. In the case of the 'light spray', 501, it is suggested that a concentration of formative forces related to light is involved as well as the physical properties of millions of tiny crystalline structures concentrating light. Although a substantial body of evidence points to the superior physical and chemical properties of Biodynamic soils, the potentially equally important role of enhanced formative forces in Biodynamic soils and plants (which may bring improved health and vitality to animal and human consumers) cannot be easily dismissed. Scientists should keep an open mind and continue to investigate this hitherto neglected field. I refer interested readers to my essay 'Formative Forces and Science' in *Biodynamic Growing*.[13]

Anecdotal Evidence

Anecdotal evidence is much less persuasive than careful scientific studies. However, it can still be illuminating and can also suggest further research possibilities. The following observations (many relating to drought conditions) were recorded by experienced Biodynamic farmers and advisers visiting Australian (and one Italian) Demeter Biodynamic farms over the years:

Despite abnormal rains and floods in the early part of the year, the farm has progressed well, but little or no rain followed, most pasture was lost. Any crops that have germinated and any pasture which survived has an amazing root development up to 4 to 6 inches in depth. This compared with the neighbour's crop and pasture 2 inches deep and spreading outwards.

Considering the late start and very little rain to the beginning of the sowing season, most of the crops are developing well. Soil has improved, comparing it with the neighbour's, which is very hard and compacted.

Even with a late start to the season and then very wet, this farm is coming away very well. In observation, this farm is progressing a lot quicker than the conventional farms around it. Cattle are in good condition with the extra feed that has developed.

Drought bad, but farm (tree lines!) stands out in the district. Colour of dry plants shine, neighbours' grey death.

Two years drought. This BD farm is standing up better than the conventional farms around here.

Drought severe. Stock very thin, but coat still shining (never seen before). Stock agent said the same!

[In a butcher's shop:] The Demeter meats on display are easily recognised by their colour, juiciness and meat texture (of relatively fine fibrousness).

Pasture is a 'glow green' colour and has active look. Good feed reserves beyond any neighbour with same stocking rate.

Even in this dry year, this farm stands out in the neighbourhood. A surprising lot of green grass. Stock in good condition.

Severe drought. Very dry, but there is a new green growth of approximately 3 inches amongst the dry grasses, which is a true BD green. The neighbours have no visible green at all, just a dead brown.

Has cattle which are fat, ready for market. I have not seen any other cattle in the area as good. There is some moisture in the soil; it will hold together when squeezed in the hand. I dug some in the neighbours' and it is like rock with no moisture in it and like dust when broken.

Drought. Pastures are very short with some new green shoots from the base of the dry clumps. There is no sign of green on the neighbours'. Cattle are in strong condition and cows near calving. Rain is desperately needed.

Started BD in 2005 from organic organization. Soil had terrible structure when started.

Farm still fully productive in a drought time. Flats have growth and bush grazing still available.

Even though this farm has come through a severe drought, it has shown considerable improvement since the last visit. Since 6 inches of rain fell 3 weeks ago, it has sprung back with legumes coming up everywhere and trees have a beautiful BD colour. This farm stands out among the surrounding farms.

Even though there is a drought and shortage of water, the cattle and crop looked in excellent condition.

A look at the neighbour's farm showed how well this farm bounced back after the rain.

The neighbour has large areas eaten by army worms whereas this farm has not been touched.

The farm is in wonderful condition and is head and shoulders above the neighbour's chemical farm.

Three-and-a-half inch downpour, all absorbed, with no run-off!

He is quite pleased at the ability of his farm to perform under such terrible, tough drought conditions. Sheep are in very good condition. Some close neighbours looked terrible.

This farm plus other BD farms in the area have recovered well, considering drought. Much better than conventional farms where wind erosion has occurred.

The best BD chocolate converted (originally white sand!) soil of all in the area.

Big production of fruit without watering (in drought).

It was commented (by the cheesemaker) that any BD cheese made as a trial was having less waste than that from the conventional milk, which makes the cheesemaker happy.

Soil very good, creek still running (August, drought). April, drought, creek still running and has fish! Quite incredible, considering. [This creek had not run all year round, or in drought, for over 80 years, but had recommended running continuously after a number of years of BD on a substantial size farm bordering the creek. Old locals are amazed.]

A dry spell was experienced after some crop had been planted and pasture germination, but no signs of wilting. Moisture retention is very good as well as organic matter. He is pleased with the results, which are showing in the cows and calves. All areas are covered with pasture which will help with wind blow, which occurs regularly in the area.

Plant colour given extreme wet indicates good drainage despite heavy black clay. This indicates that BD is beginning to work.

This horticultural area suffered extreme 48°C temperatures last summer. The BD crops were far less affected than the conventional neighbours'. The capacity of the BD trees to survive (conventional growers sustained substantial losses) extreme heat wave further demonstrated the results of BD farming. The neighbours have expressed their amazement with the yields achieved with few inputs.

Poor sandy to blue clay soils have, over a decade, developed Biodynamically to highly productive humus soils. Blue clay, possibly the most difficult of soils, no longer exists!

Dry grass has golden glow compared with neighbours, who have to feed heavily, whereas the golden glow still has food value. Neighbours' grey colour grass, need to feed reserves early.

This farm is the only one around the area that has any feed to speak of, compared with the neighbours.

Lucerne sown on recovering salty areas is doing very well, with a good root development, depth down to 10–12 inches, leaf very healthy and no sign of salt effect.

The most noticeable improvement on this farm is the improvement in coverage on the salt-affected area, which is almost covered with various plant species.

Comments have been made to him by some of his neighbouring orchardists about the longer shelf life of his fruit, which makes him pleased.

The packers and cool store personnel commented on the quality of the BD meat as well as the extra shelf life.

The farm responds to rainfall much quicker than their conventional neighbours' farms and the grass holds on longer. The sheep on the conventional block (yet to be converted to BD) require supplementary feeding and are hard to maintain in condition whereas the sheep on the BD block are 'mud-fat'. The crops on the conventional block have yielded only half as much as those on the BD farm this year. They only started supplementary feeding (drought time) on the BD block in early February, while the neighbours have been feeding for months.

When we were dairying conventionally, bloat was a real danger and losses were not uncommon. After a few years of BD farming, it disappeared completely. Nitrate poisoning also disappeared.

After 6 years of organic growing, the sandy market garden soil still had only 0.6% organic matter, the same as when they started. After six years applying the Australian Demeter Biodynamic method, the organic matter in the soil had increased to 4.2%. [Italian market garden]

Fertility problems affected 40% of cows when farming conventionally. Under BD, a 99–100% pregnancy success rate. The cows live and produce up to 6 years longer under BD. Milking 250 cows, there were calving difficulties in 50 calvings and vet visits were required in 25 of those. Under BD we had ten calving problems a year and the vet was only required once or twice a year.

After 25 years of BD, soil tests were done – the pasture roots went down 1,500mm compared with 150mm previously. The soil held ten times as much carbon as before (i.e. ten times as much CO_2 was absorbed and held as before). Before BD, irrigation water wouldn't sink in; it went stinky on top and held back the grass. Irrigation every 7 days. Under BD – water went straight in, grass grew straight away and irrigation was extended to every 30 days!

FOOTNOTES FOR CHAPTER 4

1. Mathematician and physicist, Professor of Astronomy at Cambridge from 1948 to 1973.

2. Fred Hoyle and Raymond Arthur Littleton, 'The Internal Constitution of the Stars', *Occasional Notes of the Royal Astronomical Society* (1948).

3. Rudolf Steiner advised that 500 should be stirred to form a deep vortex, then the stirring direction reversed, creating a vigorous bubbling chaos, followed by another vortex, reversing direction again, and so on for 1 hour, to prepare the 500 for spraying.

4. Correspondence between Victorian Department of Agriculture soil scientist Graeme Savage and the Extension Director at the Agriculture Department's Ellinbank Dairy Research Institute (and Biodynamic dairy farmer), Peter Medling; also correspondence between Medling and Victorian Agriculture Department Manager, Organic Farming, Ross Clarke.

5. J.A. Lytton-Hitchins, A.J. Koppi and A.B. McBratney, 'The Soil Condition of Adjacent Bio-dynamic and Conventionally Managed Dairy Pastures in Victoria', *Australia in Soil Use and Management*, 10 (1994), p. 79.

6. C.B. Parker, 'The Phosphorus Balance of a Conventional and a Biodynamic Dairy Farm', undergraduate thesis, LaTrobe University, 1992.

7. S. Cock, 'A Comparison of Soil and Plant Root Characteristics in Irrigated Summer Pasture from Two Different Farming Systems', undergraduate thesis, LaTrobe University, 1991.

8. E. Frescher and J. Russell, 'A Comparison of Biodynamic and Conventionally Managed Soils – Hyden, Western Australia', *Biodynamic Growing*, 4 (2005), pp. 29–33.

9. His results were published in a series of books including *About the Origin of Inorganic Substances*, 1873.

10. C = carbon, Ca = calcium, Cl = chlorine, F = fluorine, Fe = iron, H = hydrogen, K = potassium, Li = lithium, Mg = magnesium, Mn = manganese, N = nitrogen, Na = sodium, O = oxygen, P = phosphorus, S = sulphur, Si = silicon.

11. For more examples, see C. Louis Kervran, *Biological Transmutations*, Happiness Press, Magalia, California, 1980.

12. As outlined by Kervran in *Biological Transmutations*.

13. John Bradshaw, 'Formative Forces and Science', *Biodynamic Growing*, 22 (2014), pp. 27–33.

Chapter 5

Why Farm Biodynamically?

Why does anyone choose to be a farmer? For me, farming has been a lifelong attraction. I was drawn to nature and working in the outdoors, the physicality, the creativeness, the independence, the joy of working with living plants and animals, and, ultimately, the fulfilment of producing healthy food for people. I am sure many other farmers would share my sentiments.

I believe that, of all professions, farming is the most important. Without food we cannot live; without farmers there can be no settled society. No great cities can exist. Farmers, providing food for society, allow others to specialise in their chosen professions, safe in the knowledge that their food requirements will be met. Furthermore, farmers producing healthy food are the original doctors. Healthy food has the capacity to build robust health in consumers.

Why Organic/Biodynamic?

I would turn that question around and ask why any farmer would choose to farm conventionally in view of the history of agriculture outlined in Chapter 3. The use of water-soluble fertilisers, though in some cases producing higher yields, takes the farm away from nature's organisation, indiscriminately feeding plants. This results in

plants that are lacking in flavour and keeping quality and are more susceptible to attack by pests and diseases. The plants must then be sprayed with pesticides and fungicides to protect them. Herbicides are used to deal with weeds. Although production can be higher (though often it is not), it comes at a high cost in terms of the chemicals required and in the veterinary and other costs associated with animal health. Conventional agriculture steadily degrades soils whereas organic and, particularly, Biodynamic agriculture continually develop soils of greater depth, structure and fertility, free of chemical residues. The future of agriculture depends on, at the minimum, maintaining and, preferably, improving soil fertility and health. Biodynamic farmers know that the soils they pass on to the next generation are in a much better state than when they started, and that these soils will provide a strong environmental and economic foundation for the future, something to look back on with satisfaction and pride after a long farming career.

Health

Farmers have a responsibility to the health of not just their own families (though this is surely a matter of deep and immediate concern to all farmers) but, more broadly, of all the people they feed. Government regulators tirelessly assure us that agricultural chemicals have been through a rigorous scientific safety-testing regime and are perfectly safe for consumers and the environment. Unfortunately, the testing process is flawed. Firstly, most safety testing is done by the chemical producers themselves. Companies producing the chemicals have a vested interest in proving them safe, and situations have occurred where the profit motive appears to have overshadowed scientific objectivity. It appears that studies are at times deliberately slanted to prove safety and to minimise the chance of any adverse results. Adverse findings are sometimes obscured, minimised or omitted altogether. Many agricultural chemicals have been used for decades before it has finally been established that they are actually quite dangerous and they have been withdrawn from use. The most widely used agricultural chemical, the herbicide glyphosate, has been implicated in widespread harm to human health. Chemical companies involved have repeatedly been accused of involvement in flawed research, cover-ups and outright corruption.

The World Health Organization's International Agency for Research on Cancer reviewed approximately 1,000 studies on glyphosate and in 2015 declared it a probable carcinogen (also strongly suggesting its genotoxicity). Since then the German company Bayer (which bought Monsanto, the original maker of Roundup) has paid US$10 billion to 95,000 class action participants who claimed that their cancers were caused by exposure to glyphosate. More cases are still underway worldwide. And yet many government regulators still assert that glyphosate is safe. And farmers continue to use it, believing it to be indispensable to the control of weeds, despite the fact that farmers found ways to control weeds naturally, from the dawn of agriculture until 1974, when Roundup was first used.

One of the most serious flaws in the scientific approval process for chemicals is that the possible synergistic effects of the hundreds of agricultural chemicals we consume in our food are almost never investigated. Chemicals are safety tested by themselves despite being present in our food in combination with hundreds of other chemicals. Furthermore, evidence is emerging that some chemicals can actually be more dangerous at extremely low environmental concentrations, especially in combination with other chemicals. The validity of safety-testing regimes is coming into question.

Environmental Considerations

Farmers have a responsibility to the environment. Activities on any farm impact on a much wider area. Farming systems that predominate in an area have a cumulative effect, whether positive or negative, on whole ecosystems over broad areas.

Greenhouse gas reduction

The elevated soil organic matter levels on Biodynamic farms capture and hold carbon dioxide (1,614 tonnes per hectare in the first 6 years of Biodynamics on Alex Podolinsky's farm at Powelltown –see p. 22), thus greatly contributing to the reduction of greenhouse gases.

A European study published in 2020 (see p. 23) compared nitrous oxide emissions[1] from conventional, organic and Biodynamic production units and concluded that the organic farms produced 25% less nitrous oxide (N_2O) than the conventional ones, the Biodynamic farms producing 43% less.

Water

Biodynamic agriculture does not burden the environment with water-soluble nutrient run-off (which causes toxic blue-green algal blooms in waterways). The NPK run-off from Biodynamic dairy farms in Victoria was measured to be 30 times less than that from neighbouring conventional dairy farms in the 1991 Victorian Agriculture Department studies (see p. 22).

Biodynamic farms, with their much more open structure and greatly increased organic matter (and colloidal humus levels), absorb water far faster and in greater quantity than conventional soils. This water is held in the soil colloidal humus and is available to plants for extended periods. This results in much more efficient use of water and allows greater production in dry periods. Irrigation intervals are greatly extended. Restoration of open soil structure and increased humus levels heal stream and river systems as the delicate capillary networks that extend far back from the watercourses are restored to natural functioning and fed slowly by water released from colloidal humus. Soils that quickly absorb and hold large volumes of water not only store water for drier times but also significantly reduce the severity of flooding. The critical role played by the reduction in soil structure and humus levels caused by conventional farming, in the increasing severity of flood events worldwide, has been largely overlooked.

Salinity

Biodynamic farms have demonstrated successful, ongoing redemption of salt-affected land,[2] owing to the friable, open soil structure developed over the years. This open structure breaks the cycle of salt-laden water being drawn up from the subsoil by capillary action, allowing salt already in the topsoil to gradually move down below the root zone. Salinity redemption under Biodynamic management on the 4,800-acre (1,940-hectare) West Australian grain and sheep property of John and Bernadette Cashmore is shown in aerial photos published in *Biodynamic Growing* in 2004.[3] Scan the QR code to access the photos.

https://biodynamicgrower.com/wp-content/uploads/2024/06/Salinity.pdf

Economic Considerations

Economic issues are at the heart of many farmers' hesitation about moving to organic or Biodynamic agriculture. Many farmers feel uneasy about their use of artificial fertilisers and chemical sprays, but don't see an economically viable alternative. First and foremost, farmers must be able to make a living.

There has been a lack of research comparing the financial results of Biodynamic, organic and conventional farms. The Victorian Agriculture Department's study (see Chapter 4) comparing ten matched pairs of Biodynamic and conventional farms did do some analysis, finding higher production on the conventional farms, but higher costs in terms of fertiliser, agricultural chemicals and much higher vet bills. The net income was slightly higher for the conventional farms, but it was noted that, if environmental impacts had been taken into consideration, the result would have been in favour of the Biodynamic farms. Additionally, some of the ten Biodynamic farms studied reported much higher payment levels for their milk than their conventional counterparts were receiving (owing to their milk being sold as certified Biodynamic milk), but the final report recorded their payments as the same, a serious flaw.

Many factors are involved in the financial success of any farming operation. These include the innate quality of the soil, climatic reliability and the skill and professionalism of the farmer. These factors apply whether one is farming conventionally, organically or Biodynamically. Studies comparing the financial outcomes of different farming systems must be carefully designed and completely objective. Those conducting the studies must have at least a basic understanding of how each system works. Otherwise fundamental errors can occur, such as calculating a presumed loss of NPK from the farm solely on the basis of the amount of produce exported, irrespective of soil test results and ignoring the elevated biological functioning of soils on Biodynamic farms.

In the current absence of sufficient research on comparative financial outcomes, we must rely on the lived experience of farmers. From my experience of many visits to Biodynamic farms and many conversations with Biodynamic farmers, I offer the following observations, but stress the importance of farmers doing their own investigations.

Grain production, with or without associated sheep or cattle grazing

Many Biodynamic farmers report production levels similar to those of conventional neighbours who use moderate levels of fertiliser. Conventional growers who apply high levels of fertiliser can produce considerably more per hectare. Higher production is reported on Biodynamic farms in low-rainfall periods; higher organic matter and colloidal humus levels and more open soil structure allow faster infiltration of water and higher water-holding capacity in the soil, which sometimes makes the difference between some crop and no crop in a dry season. The longer soil preparation time on Biodynamic farms and the fact that herbicide use can allow a crop to be sown earlier than on a Biodynamic farm can give an advantage to the conventional farmer in some seasons. However, overall, with much lower costs in crop establishment and maintenance on Biodynamic farms (much lower fertiliser inputs are required and no chemical use), the net income is, anecdotally, often higher on the Biodynamic farms, even if all produce is sold on the conventional market at the same prices. Where grains and/or livestock can be sold as certified produce, the premiums obtained make the Biodynamic financial outcome even more striking. High-input conventional farmers can show higher net income in good seasons, but their high input costs can result in serious financial losses in bad seasons.

Grazing

Production levels on Biodynamic and conventional properties are often similar over time. Biodynamic properties tend to outperform conventional neighbours under adverse conditions. Grass stays green longer in dry periods; even dry grasses have more substance and can sustain animals in good condition while conventional neighbours need to use feed reserves more quickly. Selling on the conventional market, Biodynamic producers can generally expect similar net returns. Where a market can be found or developed for certified meat (e.g. certified meat wholesalers, certified butchers, farmers' markets, home delivery), much higher returns can be expected.

Dairying

The Victorian Agriculture Department, in its comparison of ten matched Biodynamic and conventional irrigation dairy farms in the Goulburn Valley, concluded that production was substantially higher on the conventional farms. This, however, was obtained at considerable extra cost, including a 600% higher level of grain feeding and frequent high-level fertiliser applications (the Biodynamic farmers were applying no fertilisers). The cows were healthier and more fertile on the Biodynamic farms, so their vet bills were significantly lower. The end result was a slightly higher net income on the conventional farms. Where Biodynamic farmers can sell their milk certified, at a premium, they can expect a considerable financial advantage. Biodynamic

dairy farmers who invest in a dairy-processing operation on-farm, to make yoghurt, cheese, butter, etc. can also obtain excellent returns by selling certified dairy products. For instance, provided markets can be found or developed, certified yoghurt products can, before costs, return wholesale, up to A\$4 per litre of milk produced.

Egg production

Comparing a Biodynamic free-range egg enterprise with a conventional free-range egg enterprise is not straightforward: under Australian Consumer Law, 'free-range' egg producers are allowed to run up to 10,000 birds per hectare, or one bird per square metre. This is not free range as far as the organic industry or consumers are concerned. Biodynamic (and organic) egg producers ensure that hens move to fresh ground regularly and always have access to fresh grass. Certified Biodynamic free-range eggs attract a healthy wholesale return, and much higher returns are possible with direct marketing (e.g. at farmers' markets). Biodynamic egg producers tend to run laying flocks of anywhere from 400 to 4,000 birds and, selling wholesale, can generate profits of around A\$1 per hen per week (where cows are milked to provide fermented milk for the hens).

Market gardening

Biodynamic market gardeners often have production levels similar to those of their conventional colleagues. Tonnages per acre will be a bit higher on conventional properties, but this extra weight is mostly extra water and mineral salts in the produce, which reduces flavour and keeping quality, as well as making the produce more prone to insect and disease problems. This in turn necessitates extra expense for remedial chemical applications. Most Biodynamic market gardeners sell their produce certified to organic/Biodynamic wholesalers or direct at farmers' markets or similar, obtaining higher prices than their conventional colleagues. Generally, slightly lower production levels on Biodynamic market gardens will be offset by higher returns, whether selling wholesale or retail. Net returns can be considerably higher on Biodynamic properties.

Fruit production

Biodynamic fruit growers tend to have production levels similar to those of their conventional counterparts. As with market gardeners, any extra weight of conventional crop consists of extra water and salts, with resultant lowering of flavour and shelf life. Conventional growers have much higher costs in terms of fertiliser inputs and chemical sprays. In very adverse, particularly excessively wet, climatic conditions, a skilled Biodynamic operator can often still produce healthy fruit when no amount of chemical use can save a conventional crop. Biodynamic growers generally sell their fruit certified, gaining higher prices than conventional growers.

These observations are just that – observations – and don't have the authority of carefully and objectively designed studies. However, they are based on many discussions with Biodynamic growers all over Australia and on general industry knowledge. Individual growers are advised to do their own investigations:

- talk to Biodynamic and organic producers;
- visit farmers' markets, health food shops and certified butchers and talk with meat wholesalers;
- look at how other certified producers are marketing their produce;
- consider value-adding – for example, dairy processing, dried fruit, juice production, herb teas, olive oil processing, flour milling, pasta production, organising butchering and packaging of meat;
- for Australian producers, consult the Biodynamic Marketing Company (BDMC), the not-for-profit company set up to assist in the wholesale marketing of certified Biodynamic and organic produce in Australia (BDMC can provide insightful advice on opportunities and limitations in marketing certified Biodynamic produce).

FOOTNOTES FOR CHAPTER 5

1. Nitrous oxide constitutes 7% of all greenhouse gas emissions from human activity and is the biggest threat to the ozone layer.
2. See Alex Podolinsky, 'Salinity', *Biodynamic Growing*, 3 (2004), pp. 19–24.
3. Ibid., pp. 21, 24.

Chapter 6

Biodynamic Soil Development

One of the great joys of practising Biodynamics is to see the progressive development of one's soil. A cold, unstructured, lifeless soil gradually darkens and comes alive, teeming with microbes, worms and other soil life, and becomes crumbly and structured. Terms such as 'warm, glowing, organised structure' begin to describe what one sees. The increasingly healthy, vibrant, upright plants reflect this soil development.

The open structure, filled with air pockets, brings not only improved drainage but also improved water infiltration and water-holding capacity (increasing humus levels hold more water). Air is the most important element in soil. Soils that 'breathe' support aerobic bacteria and other micro- and macro-life that are so essential to the healthy biological functioning of soil.

Livestock show a marked preference for Biodynamic pasture, even a few weeks after the very first 500 spray. Problems arose in the early years of Australian Biodynamic development when some farmers applied 500 to only part of their farm as a trial: animals would refuse to graze the unsprayed paddocks, waiting at gateways to be allowed back into the sprayed paddocks.

Assessing Soil Structure

How can we assess our soil structure? What is good soil structure and what is bad soil structure? Imagine you are baking a cake and have forgotten to add the baking powder: the cake is heavy and compacted and has no air pockets. With the baking powder, it is light and fluffy and has many air pockets. This is the difference between poorly structured and well-structured soil; air pockets are of fundamental importance. A well-structured soil develops a rubbly crumb structure, with lumps of varying sizes held together by fine hair feeder roots, and contains many air pockets. There is organisation in the structure, as in a well-designed building. Hand in hand with the developing rubble structure comes a deepening and enriching of the soil colour as the organic matter and humus levels increase. The soil breaks along natural fracture lines, not into layers of semi-compressed material. The soil gradually looks warmer – and actually it *does* become warmer.[1]

Assessing soil structure is a simple process. Dig into the soil with a garden fork, shovel or spade, preferably to its full depth. This is usually only possible when there is some moisture in the soil, though not impossible if the soil is particularly well structured. Slowly pull the handle of the fork backwards to break open the soil. We want the front surface to break out naturally and be exposed for viewing.

soil by digging around all four sides will show nothing of the soil structure. It must be broken out naturally.

Figure 6.1.

Figure 6.2.

Figure 6.3. Soil opened with fork, broken out naturally. Here we can see a very nice open, rubbly structure developing in the top 100 mm along with deepening colour. Less structure below and lighter colour. Fine hair feeder roots and some mycorrhizal fungi (the white areas).

Developing Soil Structure

How does one bring about Biodynamic soil conversion and, in all one's interactions with the soil, protect and further develop the emerging structure? There are many factors that must be considered and correctly applied if Biodynamics is to work. Each factor that is not right will reduce the effectiveness of the Biodynamic preparations. If enough factors are not right, there may be no result whatsoever.

Depending on the soil, it often helps to place your foot on the soil in front of the fork and apply light pressure as you lean back on the fork. This helps ensure that the front face of the soil spit will break out cleanly. The soil spit can be examined in situ (easier to replace) or removed for examination if desired. Extracting a spit of

Every single development in the Australian Demeter Biodynamic method has been rigorously tested, in the laboratory and in the field. The following points all result from this scientific process and have been proven for all climates and on many continents. They are not hard to follow, but each point is important for the successful development of your soil. Just as in learning a musical instrument there are basic techniques for the care of the instrument, playing techniques, learning of scales, etc., these Biodynamic pointers are some of the basic essentials. Once they have been mastered and applied correctly, your soil will develop successfully. Later, your individual observation skills and experience will enable you to become objectively creative in your own unique situation, but always with a foundation of proven correct practice. A more detailed description of each point can be found in Part 2.

Using high-quality Biodynamic preparations

High-quality preparations are central to Biodynamics. The art and science of preparation-making are well established. Years of experience are required, as are the highest-quality raw materials, a well-sited and well-prepared pit and a cold winter climate. It is most important that preparations 500, 502, 503, 504 and 506 should be in a moist and colloidal condition when retrieved from the soil and in storage. Preparation 505 (oak bark) should be kept moist but is not a colloidal preparation, while 501 is a dry powder and 507 is a liquid.

Correct storage of the preparations

A well-made 500 storage box (and compost preparation storage box if needed, for frequent large-scale composting, as described on p. 99) and regular inspection of the 500 (and compost preparations if stored), adding drops of water as

required, ensure that the preparations remain as moist and colloidal, and therefore effective, as when received from the Biodynamic preparation supplier.

Correct application of 500 and 501

This includes: proper heating, preferably as a whole, of the purest water available to 28–33ºC for the larger (20-acre) capacity stirring machines, which hold warmth for longer, or up to 35ºC (never higher than 35ºC) for 8-acre machines or smaller set-ups; a well-tuned stirring machine that creates a deep, vigorous, straight-sided vortex and vigorous, bubbling chaos; and an efficient spray rig to ensure low-pressure, even distribution of large droplets (and high pressure application of 501 as a mist), applied when weather, soil, pasture and crop stage are suitable.

Developing humus in the soil

This occurs through the use of prepared 500, the addition of Biodynamic colloidal humus compost where practicable, the growing and incorporation of green manure crops, and the correct management of pastures, together with harrowing after cattle when conditions suit. Many of these practices can be grouped together under the heading 'Biodynamic sheet composting' (see Chapter 17).

Careful cultivation

Choices of implement, timing and speed of cultivation all play an important role in conserving existing soil structure and further developing it. Cultivation is covered in detail in Chapter 16.

FOOTNOTE FOR CHAPTER 6

1. See Frescher and Russell, 'A Comparison of Biodynamic and Conventionally Managed Soils', p. 33.

Figure 6.4. Soil in avocado grove, New South Wales, after 4 years of Demeter Biodynamics.

Figure 6.5. Soil in neighbour's conventional avocado grove.

Figure 6.6. Compacted sandy clay loam soil. Reproduced from the Podolinsky Collection with kind permission.

Figure 6.7. Same soil as in Figure 6.6 after 1 year of Biodynamic management. Reproduced from the Podolinsky Collection with kind permission.

Figure 6.8. Soil conversion on a low-rainfall (325mm) Western Australian Biodynamic farm. The original white sand is at the bottom. Before Biodynamics these white sands would blow if cultivated. Now they can be safely cultivated.

Figure 6.9. Biodynamic conversion of very heavy, difficult-to-work, gimlet clay soil, on the same farm as Figure 6.8. Originally, it was impossible to grow good grain crops or pastures on these heavy soils, but they can now support good crops and pastures.

Figure 6.10. Australian Demeter Biodynamic method vineyard in Bourgogne, France, in which Marc Guillemot has deep ripped a row. Photo: Guillemot family.

Figure 6.11. Marc also deep ripped a row in his neighbour's vineyard. Photo: Guillemot family.

Figure 6.12. Guillemot vineyard soil before conversion to Biodynamics. Photo: Guillemot family.

Figure 6.13. Guillemot vineyard soil after conversion to Biodynamics. Photo: Guillemot family.

Figure 6.14. Conventional neighbour's soil (left) and the Guillemots' soil (right). Photo: Guillemot family.

Figure 6.15. Biodynamic pasture, northern Victoria, no rain for 3 months. Reproduced from the Podolinsky Collection with kind permission.

https://biodynamicgrower.com/wp-content/uploads/
2024/01/Biodynamic-Soil-Additional.pdf

Figure 6.16. Neighbour's conventional soil, no rain for 3 months. Reproduced from the Podolinsky Collection with kind permission.

Figure 6.17. Soil conversion on the author's clay loam soil after 2 years of prepared 500 application.

Figure 6.18. The same soil as in Figure 6.17 after 5 years of prepared 500 application.

Chapter 7

Biodynamic Upper Plant Expression

Where Biodynamics is properly applied, using high-quality Biodynamic preparations, together with other Biodynamic practices, certain results can be expected. Soil biological activity will greatly increase together with soil humus levels and root activity. This will result in the development of open, friable soil structure, which will progressively extend deeper. Plants growing in these soils, and feeding from soil humus, within nature's organisation will exhibit the following observable characteristics, which will be accompanied by beautiful, sweet flavours and extended keeping qualities:

- Plants will have a more upright, vibrant, more refined form.
- Leaves will be well placed, in a more organised, purposeful and harmonious arrangement.
- Root structures will be extensive and well organised, penetrating deeper and with abundant fine hair feeder roots.
- Pastures and crops, when viewed at their best, exhibit a Biodynamic glow, as if lit from within. This is called 'glow green' by experienced Biodynamic farmers.
- Plants will be stronger and more resistant to insect and disease problems. When

adverse weather conditions prevail, such as sudden cold snaps, severe storms, drought or excess moisture, plants will become more susceptible to problems, but many situations can be restored to balance by timely use of, for example, 501 or valerian.

It is important for Biodynamic growers to steadily develop their powers of observation in nature. Farmers, living and working in nature, are generally far more observant than the average city dweller. Their eyes are always keenly attuned to the weather, soil conditions, plant growth and health and the wellbeing of their animals. Biodynamic growers can further develop this observational ability, learning to actively engage with nature and view things as a whole. This deeply engaged, holistic, active viewing can, for instance, tell a farmer whether a crop or plant is growing healthily and has a vibrant, upright expression, or whether it is overblown, or stagnating, and in this case enable the farmer to take necessary steps to redress the imbalance before it causes problems.

Such active viewing has commonly been exercised by people living close to nature. Many farmers have this ability without being conscious of it; it brings valuable insights for farm decision-

Figure 7.2. Older Biodynamic eggplant. Note the upright expression and harmonious placement of the leaves.

Figure 7.1. Young Biodynamic eggplant. It is in balance, having enough colloidal humus to grow strongly, but is not overfed. The upper plant expression is upright and vibrant.

making. Active viewing relies on a calm and concentrated 'extending out' through the eyes. Excessive viewing of television, video games or mobile phones can seriously damage the ability to actively view, especially when we are often exposed to rapidly changing images on a screen. The damage caused to active sight by the increasing use of rapidly changing images cannot be overstated. As generations of children exposed to such images in video games grow up and become producers of movies and advertising, the situation becomes worse and worse, to the point where many older people (and those who avoid watching such images or playing video games) can no longer stand to watch these images. They are (often unconsciously) avoiding watching such images because it pains them to do so. They are protecting their ability to view in nature.

Alex Podolinsky developed a method of consciously training our active viewing capacity, calling it 'active perception'.[1] He trained many people worldwide, including teachers, doctors, intellectual disability workers, artists, architects and, especially, farmers, in the method. Those interested can read (and practise) *Active Perception* as well as material based on Podolinsky's method, such as Maria Bienert's three 'Art and Farming' articles[2] and my essay 'Aspects of Form',[3] which explores the influences of earth, water, light and warmth in trees as introduced in Podolinsky's *Biodynamic Agriculture Introductory Lectures Volume 1.*

Active perception training is beyond the scope of this volume. However, Figures 7.1 to 7.13 give a sense of what Biodynamics can achieve in upper plant expression, contrasted, in some cases, with conventionally grown and overblown organically grown plants.

*Figure 7.3. Organic beans (Europe).
Photo: Maria Bienert.*

*Figure 7.4. Biodynamic beans (Australian Demeter
method, Europe). Photo: Maria Bienert.*

*Figure 7.5. Biodynamic beans. Alive, upright and
energetic upper plant expression.*

FOOTNOTES FOR CHAPTER 7

1. Alex Podolinsky, *Active Perception*, Gavemer, Sydney, 1990.

2. Maria Bienert, 'Art and Farming', *Biodynamic Growing*, 8
    (2007), pp. 16–22; 9 (2007), pp. 17–22; 18 (2012), pp. 12–18.

3. John Bradshaw, 'Aspects of Form', *Biodynamic Growing*, 1
    (2003), pp. 31–33.

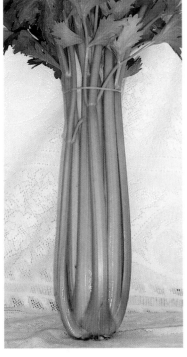 

Figure 7.7. Organic 'overblown' celery, pumped up by water soluble elements, likely resulting from the use of pelletised manures or non-colloidal compost. Very similar in appearance to conventional, artificially fertilised celery.

Figure 7.6. Biodynamic celery grown in a sandy soil with a generous Biodynamic colloidal humus compost application before transplanting.

Figure 7.8. Biodynamic greenhouse cucumbers – upright expression, 'dancing' leaves. Photo: Maria Bienert.

Figure 7.9. Heavy, over-fertilised organic cucumbers. Photo: Maria Bienert.

Figure 7.10. Bluey-green, heavy, artificially fertilised conventional broccoli. The leaves flop down. Photo: Maria Bienert.

Figure 7.11. Biodynamic broccoli with healthy colour and upright leaves. Photo: Maria Bienert.

Figure 7.12. Conventional vineyard, Saverio Petrilli's neighbour, Italy. Photo: Saverio Petrilli.

Figure 7.13. Saverio Petrilli's Biodynamic vineyard, Italy. Photo: Saverio Petrilli.

Chapter 8
Balance in Biodynamics

The fundamental Biodynamic techniques, established by scientific experiment and extensive field trials, provide a proven framework for the farmer to work within. However, within this framework, Biodynamic farmers have to develop their own observational skills and creativity. They learn to read nature and act according to the needs of their farm at any given time.

A simple example. We know that the harrowing of cow manure after cows have grazed a paddock is very beneficial: the manure is spread thinly over the pasture to avoid an excess of water-soluble elements in the area of a cow pat and the wasting of grass when the cows avoid eating the overblown grass around the cow pat, and at the same time the valuable effects of the manure are widely spread, enabling the soil life to quickly incorporate the manure into the soil humus. If the farmer is not awake and observing nature, they may continue harrowing paddocks too long into winter (or whatever is the wet season in their region), not noticing that the tractor wheels are compacting the wet soil. Being mindful and observant, a good farmer will stop harrowing when the wheels start to make noticeable tracks in the pasture, meaning that soil compaction is occurring, and not start again until it is safe to do so in spring. On the other hand, if the winter is dry, harrowing can continue, to much advantage.

Biodynamics is full of such considerations and the individual farmer is constantly challenged to be mindful and observant and make decisions according to the conditions that are presented. Often, action can be taken to balance a situation that is unbalanced. Some of the main areas where balance is required are as follows.

Soil Type

Soils vary enormously according to the proportion of sand and clay they contain. Very light, sandy soils lack fertility but are easy to cultivate, and they drain well. Heavy clays can be very hard to cultivate and don't drain well. The ideal is a blend of clay and sand, in varying degrees of sandy loam, loam or clay loam. On a farm scale, it takes time to balance a predominantly sandy or clay soil. Sands are improved by regular seasonal spraying of prepared 500, which increases microbial activity, root growth and humus formation, the soil gradually developing into a darker, well-structured sandy loam. This can occur more quickly in a higher-rainfall area. Alex Podolinsky dramatically transformed the poor sandy soil on his farm at Powelltown (1,500mm average annual rainfall) within six years, using 500 and the other Biodynamic techniques (see pp.17–18, 22). New soil was effectively created as the impoverished sand turned into a humus-rich, dark, fertile sandy loam.

Green manure crops can be grown to accelerate the process. Biodynamic compost, if available, can also accelerate the balancing of a sandy soil. On a garden scale, more rapid improvement can be achieved with regular green manures and compost applications in combination with prepared 500. On a very small scale, gardeners can also spread powdered clay and work it into the soil to balance its sandy nature.

Heavy clay soils will also gradually become more balanced through the use of prepared 500. For instance, Biodynamic dairy farmer Don Rathbone turned 75mm of topsoil on heavy clay into 900mm of topsoil over a 30-year period, using prepared 500 together with the other Biodynamic techniques, including deep ripping pasture in a semi-dry state, spreading Biodynamic compost (made with manure and hay from his own farm) on deep-ripped paddocks and irrigating it in, rotational grazing, pasture harrowing, etc.[1] Occasional applications of gypsum (calcium sulphate) can help by clumping clay into aggregates, thus opening up the soil temporarily.

Deep ripping when a clay soil is semi-dry will crack the soil, allowing better root penetration, which, in conjunction with 500 application, will gradually structure the soil and moderate it with increasing humus content. Green manures and Biodynamic compost applications, where appropriate or possible, will greatly assist. On a smaller, garden scale, similar measures can be used (by means of hand tools), and quicker results are possible with regular green manuring and compost applications together with careful cultivation. Gardeners also have the possibility of spreading some sand (e.g. washed river sand, as used in concrete) and gradually working it in (ideally after an application of gypsum has temporarily aggregated the clay and when it is not too wet). Thus, soils that are unbalanced by too much sand or clay can be balanced over time.

Soil Fertility

It takes time to get to know one's soil and its fertility level, and the requirements of various crops one wants to grow. Biodynamic growers learn to judge the requirements of their soil in relation to the particular crop being grown. Easiest is a poor sandy soil: in this case you know, or will very soon realise after your first attempts at growing something, that it won't sustain strong healthy growth without sufficient colloidal humus. For instance, in a light sandy soil, large applications of colloidal humus will be required to grow a successful heavy feeder crop such as celery or cauliflower. The colloidal humus can be supplied by digging in good-quality colloidal compost or by growing a green manure fed by manures, chicken manure pellets, blood and bone or similar. (For certified production, grow two green manures in a row, digging in the first then sowing a second, the second having no manures applied.) The green manure is chopped up and cultivated in around flowering time, then enough time is allowed for the green material to be digested and converted into colloidal humus before planting (often 4–6 weeks will be sufficient). A subsequent light feeder crop may do well without additional humus being supplied.

In a fertile loam soil, on the other hand, one has to be careful not to overdo the fertility by adding too much compost or too generously fertilising a green manure crop, otherwise the plants can be pushed and become overblown and watery. Balance has to be maintained.

Water, Fertility and Light

Biodynamic growers learn to observe their plants' form and structure and recognise when the upper plant expression is upright and alive and when it may become overblown, heavy or even droopy – and also recognise when the plants may be showing good upper plant expression but be suffering from insufficient elements and hence poor growth. Biodynamic plants (indeed all plants growing within nature's organisation) should obtain their elements from the soil humus, as happens in nature, not from soluble elements dissolved in the soil water or from soluble elements sprayed on the leaves.

A balance must be maintained between the soil/water side and light. Crops can become over-watery or over-lush and lack uprightness:

- During overly wet periods, or when excess irrigation water is applied.
- During extended low-light or cloudy periods, whether cold or warm (in warm, humid, cloudy periods, plants will continue to take up minerals because of the sun warmth, but lack of sunlight can result in the plants being unable to properly assimilate these minerals).
- From an over-supply of colloidal humus – adding too much compost to an already rich soil, or over-fertilising a green manure crop before a food crop.

In these conditions, plants lack flavour and nutritional quality, can become attractive to pests such as aphids and caterpillars as well as fungal diseases and have a lower shelf life after harvest. Under these conditions, we can apply the unique Biodynamic spray made from powdered quartz crystals, 501, to bring more 'light' into plants. Plants will greatly benefit from the stimulation of light metabolism/photosynthesis brought by spraying 501. Autumn-planted vegetables (e.g. brassicas, Chinese cabbage), which will continue growing under cold conditions, will also benefit immensely from an application of 501, which will balance the often low-light conditions going into winter, bringing beautiful flavour and keeping qualities. Used correctly, 501 returns plants to balance, overcoming excessive wateriness or lushness with enhanced light influence and building upright, lively plant expression, extra flavour and sweetness, and nutritional quality.

One summer I made the mistake of leaving a hose running under a lemon tree for a few hours. The ensuing crop of lemons had skin spots, and a nearby hibiscus was attacked by aphids (which had never happened before). Many gardeners have problems with aphids on their roses in spring and with caterpillars in sweetcorn and tomatoes. My experience has been that these issues are consequences of excess fertility or excess water, or a combination of the two, and are not usually a problem provided balance is maintained. If they do become a problem, 501 is the answer, combined with some natural methods of reducing the pests in the interim – for example, *Bacillus thuringiensis* for caterpillars.

Cold and Warmth

Sudden cold snaps can cause problems with crops. For example in early spring, when fruit tree buds are swelling or have already begun flowering, a sudden cold snap can cause 'sap shock', leading to possible fungal problems. If valerian preparation (507) is sprayed, preferably hours before the cold snap, but also as soon as possible afterwards, it brings a warmth influence that can help support the plant's return to balance. It has also been shown to help plant recovery after damage from severe storms.

Companion Planting

Growing together plants that help each other, and avoiding growing together plants that hinder each other, has long been practised. This practice was recognised and encouraged by Rudolf Steiner in his agricultural lectures of 1924. Many factors influence how plants grow in relation to one another: scents, insect repelling or attracting compounds, root exudations, root type and activity, physical characteristics (a tall growing plant may assist

another plant by providing shade or allowing it to climb on it, e.g. corn and climbing beans), and many more. Biodynamic practitioners have studied and recorded the effects of many plant combinations. These were described in Philbrick and Gregg's *Companion Plants* and have since been reproduced in many books on companion planting.[2] Companion planting is ideally suited to the home garden, where small amounts of a wide variety of vegetables, flowers, herbs and fruit are grown. It is less easy to incorporate where large commercial crops are grown.

If we follow the established fundamentals of Biodynamics, which have been thoroughly tested and proven in extensive field trials over time, we will achieve wonderful results in terms of soil structure and fertility, plant and animal health, flavour and keeping quality. Individual farmers are responsible for determining when and how to act, within those fundamental principles. There is much scope for individual creativity and ongoing development of the Biodynamic method, provided we continue to be balanced and retain scientific objectivity.

FOOTNOTES FOR CHAPTER 8

1. See Don Rathbone, 'Biodynamic Farming', *Biodynamic Growing*, 2 (2004), p. 19.

2. Helen Philbrick and Richard B. Gregg, *Companion Plants and How to Use Them*, Robinson & Watkins, London, 1972.

Chapter 9

Transitioning to Biodynamics

Biodynamics can and should become the agriculture of the future. It holds the answers to many problems that beset modern agriculture and the wider natural environment. It builds open, friable soils to depth, increasing humus levels and biological activity, absorbing and holding more water, reducing flood severity, healing salinity and absorbing and storing vast amounts of carbon dioxide. By bringing soils to a high state of biological functioning, it reduces the need for fertiliser inputs, thus conserving the earth's finite phosphorus reserves. Indeed, as phosphorus becomes scarcer and more expensive to extract, the economics of conventional farming will become more and more problematic.

Changing your farming system, however, does not happen overnight. It takes a certain amount of preparation and the development of a new outlook. From reading some of the Biodynamic literature in the 'Biodynamic Resources' section of this book, and talking with practising Biodynamic farmers, a new perspective will gradually develop. Real desire for change must take hold before a farmer is ready to take what, to many, will seem an alarming leap of faith.

It is advisable to do some research on the prospects for marketing your farm produce. It may be that you decide to go ahead on the basis that you want to transition to Biodynamics but still market produce uncertified, or you may identify opportunities to maximise income by becoming certified. In the early years of Biodynamic development in Australia, the majority of farmers who converted did so because they recognised the enormous benefits to the health of their farm and to their families; the demand for organic/Biodynamic food lagged behind the vast acreages converting to Biodynamics, so most Biodynamic production was sold on the open market. Now there are far more opportunities to market produce as certified, with the associated premiums.

The motivation to convert to Biodynamics (or indeed to organics) should always stem from a genuine appreciation of the benefits to the land and nature rather than solely from a desire for additional income, though that will often come as a welcome bonus.

Having done the necessary reading and research, and made the decision to transition, contact your national Biodynamic association for further help (see 'Biodynamic Resources').

The first practical step is to obtain a suitably sized stirring set-up for 500 (which can also be used for 501) and a spray rig to match (see Chapter 11). Generally, for a small property up to around 100 acres (40 hectares), an 8-acre stirring machine will suffice. Larger properties are better served with a 20-acre machine, and as the size of the property increases, multiple (20-acre) barrel machines operating in tandem, driven by the same drive system, should be obtained or built. The spray tank should match the size of the stirring set-up. As soon as weather conditions and soil moisture allow, 500 spraying can begin.

Important Considerations

- Transition from organic to Biodynamic is often accomplished more smoothly than transition from conventional to Biodynamic, depending on the length of time the farm has been run organically, the innate fertility of the soil and many other factors. Chemical use has already ceased on an organic farm and, hopefully, biological activity and humus development will already be underway. Every situation is different, however, and attention to detail in the application of the method, including the use of highest-quality Biodynamic preparations, will bring steady improvement in even the most damaged soils.
- The property should be converted as a whole. Once conditions permit, the whole farm should be sprayed with 500. If only part is done as a trial, stock will be reluctant to graze unsprayed paddocks after they have tasted the grass sprayed with 500. Wool

breaks have occurred when sheep have been rotated from sprayed to unsprayed areas and back.
- For very poor soils or ones that are highly chemicalised, straight 500 (see pp. 5–6) should be used for the first few applications before switching to prepared 500.
- Until soils have had a chance to develop biologically and humus levels to increase, some allowable fertiliser inputs may be required, reducing as the soil develops. Soil testing can be a useful aid in the early stages, together with careful observation of plant growth and animal health.
- Plants and animals may well need extra care and attention during the first few years of conversion. Many problems will gradually reduce and eventually virtually cease as soil health and vitality build. Crops or pastures that are attacked by red-legged earth mite or lucerne flea, for instance, will steadily become less susceptible. Biodynamic techniques are available to strengthen plants under such attack (see pp. 87–88). In cases where crops may be lost, a farmer may have to resort to a chemical spray in the early stages of conversion, but of course this will delay the certification process if you are seeking certification. Be prepared to drench susceptible animals for worms, preferably with allowable natural materials, especially if you are seeking certification. However, in serious cases, animal health must be prioritised and a chemical drench may be needed to save lives.
- Once regular chemical drenching ends, susceptible stock will become obvious. They

will need special care. Susceptible animals can be progressively sold off and stronger animals used to slowly develop a more worm-resistant cohort. Techniques such as running a programme of worm egg counts in ram faeces can help with worm-resistant breeding. This process can take some years.

- Soil chemical residues may be an issue. Soil testing should be done to identify any potential problems if you are seeking certification. Normally, residues of substances such as organophosphates, organochlorines and heavy metals[1] must be below certain levels for certification. When buying a new property, the contract of sale should include a clause 'pending acceptable chemical residue levels as determined by soil testing' or similar. Advice can be sought from your national Biodynamic association about acceptable levels. In some cases, the testing of produce is also relevant: although a problematic level of a chemical may be found in the soil, it may be absent from particular types of produce grown in that soil. Moreover, experience has shown that, after a number of years of Biodynamic practice, plants feeding from newly created soil and humus may test free of the contaminant. For instance, a Biodynamic asparagus producer in Victoria, whose soil contained problematic levels of breakdown products of DDT, carried out an ongoing testing programme and found that after a few years the asparagus tested clean. This is not to say that chemical contamination can always be solved quickly. On this property, marrows grown on the same land still contained breakdown products of DDT despite the asparagus testing clean.

- Generally an experienced Biodynamic farmer can convert a new piece of land more rapidly than can a newcomer. Many farmers make mistakes in the early years of their Biodynamic practice. This is only natural, but with time one's understanding and skill in application improve and the whole process accelerates.

FOOTNOTE FOR CHAPTER 9

1. Check with your national Biodynamic certification organisation for the substances normally tested for in your country.

Chapter 10

Organic and Biodynamic Certification

Most countries have organic certification systems that enable organic and Biodynamic farmers to be audited and certified if they meet the certification standards. Certification gives consumers confidence in the integrity of farmers' produce and gives farmers marketing and price advantages as well as export opportunities. Many countries shape their organic standards to meet international requirements so that their farmers' organic certification will be recognised in other countries. This is overseen by International Organic Accreditation Services (IOAS).

In Australia, there is no requirement for a farmer to be certified; some are happy to just farm organically or Biodynamically and see no particular advantage in becoming certified. However, certification gives consumers assurance of the integrity of the produce. Without this assurance, consumers may be misled by unscrupulous individuals farming conventionally but marketing their produce as organic. Some openly state 'we follow organic methods, but just use a little Roundup occasionally' – which means they cannot be engaged in organic farming.

In many settings, organic certification is mandated; for instance, organic wholesalers and many health food shops will not accept organic produce unless the producer is certified. Major Australian farmers' market associations will not allow traders to call their produce organic unless they are certified. There have been cases in Australia where traders at non-accredited farmers' markets have bought conventional produce at wholesale markets and dishonestly sold it as organic.

Some countries have more stringent requirements on anyone wanting to sell organic produce. In the United States, for instance, anyone marketing organic produce over the value of US$5,000 annually must be certified. Those marketing organic produce worth less than US$5,000 are not required to be certified but must still follow US Department of Agriculture National Organic Program (NOP) organic regulations and maintain accurate records.

Australian Organic Standards

We will now look in detail at the Australian situation, as an example. Farmers in other countries should obtain a copy of their own country's organic and Biodynamic certification standards.

In Australia, the first organic standard, the Australian Demeter Biodynamic Standard, was launched by the BDRI in 1967. The BDRI was at first denied the right to certify Biodynamic produce by the Australian Agriculture Department, which would not recognise that such produce differed in any way from conventional produce. Fortunately, Alex Podolinsky was able to convince senior bureaucrats and allowed to proceed, a major pioneering step for the whole organic industry in Australia. Biodynamic farmers could then be certified by the BDRI and could market their produce as certified Biodynamic and display the international Biodynamic Demeter logo.

The Demeter symbol is recognised worldwide as certifying the finest-quality Biodynamic produce. Demeter is the ancient Greek goddess of agriculture, a most appropriate patron. In Australia, the BDRI was founded in 1957 and was vested with the Australian rights to the Demeter trademark in 1967. It operates independently of the Biodynamic Federation Demeter International in Australia.

In many countries, the Biodynamic Federation Demeter International (formed in 2020 from Demeter International – founded in 1997 – and the International Biodynamic Association – founded in 2002) administers the International Demeter Biodynamic Standard which is used by national organisations to certify produce as Demeter Biodynamic.

The (Australian) National Standard for Organic and Bio-Dynamic Produce was implemented in 1992, allowing all organic and Biodynamic farmers to become certified and market their certified produce. The standard was developed under the supervision of the Australian Quarantine Inspection Service, a section of the federal Agriculture Department, to ensure the authenticity of organic produce exported to other countries. It was developed in conjunction with Australian organic industry bodies, which provided the professional expertise, and with compatibility with overseas organic regulations in mind. The National Standard is now overseen by the National Standards Advisory Committee (NSAC), a body within the federal Department of Agriculture, Fisheries and Forestry. The NSAC provides advice and recommendations to the department on proposed revisions and updates to the National Standard.

The five Australian organic certification organisations are the BDRI, ACO Certification Ltd, NASAA Certified Organic, Organic Approved Certifying Bodies (Organic Food Chain) and Southern Cross Certified Australia Pty Ltd.

There are two levels of certification: *in-conversion* and *fully certified*. In-conversion status can be obtained after one year of applying organic or Biodynamic methods, abiding by the National Standard, and entitles the operator to display the logo of the relevant certification body, accompanied by the words 'in-conversion'. Operators who can demonstrate that they have been abiding by the National Standard for one year or more prior to their first inspection by a certifying body may be able to obtain in-conversion status soon after the first audit by a certifying organisation. The same applies to organic producers who want to be certified organic by the BDRI, but Biodynamic in-conversion can only be granted 12 months after commencing Biodynamic practice, which includes spraying 500 on the farm.

Anyone who wants to become organically or Biodynamically certified should download and read a copy of the National Standard for Organic and Bio-Dynamic Produce.[1] The current version (at the time of printing) is Edition 3.8 November 2022. New versions are issued periodically.

The five certifying bodies each provide a certification logo to their certified operators. In addition to the Demeter Biodynamic logo provided to certified Biodynamic operators, the BDRI also provides an organic certification logo to those who become organically certified with BDRI.

Australian Demeter Bio-dynamic Standard

Biodynamic farmers who want to be certified Demeter Biodynamic must comply with both the National Standard for Organic and Bio-Dynamic Produce and the Australian Demeter Biodynamic Standard. The Australian Demeter Biodynamic Standard adds some critically important requirements, the most important being that Biodynamic in-conversion status is not granted until initial signs of soil development are apparent (evidence of the beneficial effects of the use of 500 and other Biodynamic practices), and that full certification is not granted until considerable soil structure

and humus development can be demonstrated (all being well, three years after commencement of Biodynamic farming practices). Thus Demeter Biodynamic certification requires not just the absence of artificial fertilisers or chemical sprays, but also a demonstration of positive changes to soil biology and structure, which will ultimately be expressed in higher-quality, better-tasting produce.

Demeter Biodynamic certification is available to any farmer who can demonstrate compliance with the standard. Membership of the BDAAA is not a requirement, but does provide training, mentorship and the opportunity to be a part of a welcoming and energetic community of farmers.

Certification Fees

In 2024, the annual BDRI certification fee for farmers was A$1,512 (including goods and services tax [GST]), plus 0.1% of gross farm sales. Farmers who also carry out on-farm processing pay A$1,512 annually plus 0.2% of gross sales. There are also some initial application and inspection fees. Contact BDRI for up-to-date fees or check their website: www.demeter.org.au. In addition, the operator pays for an initial comprehensive soil test to ensure there are no undesirable chemical residues in the soil.

Producers whose turnover is very low may not feel they can afford to become certified. They should consult BDRI, since there may be a possibility of reducing the annual fee.

The Application Process

Download the application form from the BDRI website. The application requires details such as any fertiliser or chemical use before commencing Biodynamic farming, an outline of your Biodynamic farming practices, the source of your Biodynamic preparations, the length of time you have been farming Biodynamically, and many other details. A farm management plan, which outlines how you run your property, is also required, together with a map showing the farm layout, including quarantine areas, named or numbered paddocks and the area in hectares of each. BDRI has pro forma management plans that assist in this.

The completed application, together with all required attachments, should be sent to the BDRI without delay. The clock is ticking: the date you become eligible for Demeter in-conversion status is one year after the whole property has been sprayed with 500 (except for any steep or otherwise inaccessible areas that can't be sprayed), but it is very important to get the application, together with the initial application fees, to BDRI as soon as possible after that initial 500 spray. If you are using Biodynamic preparations other than those supplied by the BDAAA or Biodynamic Growers Australia Incorporated (BDGAI), consult BDRI as early as possible. The quality of the preparations will need to be assessed, since poor-quality preps could delay the certification process.

Provided the application is in order, an auditor will be sent by the BDRI to conduct the initial inspection of the operation. They will conduct a thorough audit of the farm operation and also take soil samples to test for chemical residues in the soil (and sometimes also the produce). One year after the initial application,

a BDRI inspector will again visit and assess progress. Provided all is in order and the soil shows some initial Biodynamic development, the operator will be granted Demeter in-conversion certification (or organic in-conversion if the operator wants to be certified organic).

BDRI inspections are carried out annually (occasionally more often) from then on. Three years after the initial application, the operator becomes eligible for full Demeter Biodynamic certification. This is, however, not a foregone conclusion: the auditor must see considerable Biodynamic development of soil structure and humus development, together with Biodynamic upper plant expression and colour before they will recommend full Demeter Biodynamic certification. Provided the operator follows Australian Demeter Biodynamic guidelines, this soil and plant development should progress naturally.

FOOTNOTE FOR CHAPTER 10
1.   Obtainable free online at www.agriculture.gov.au

Part 2

Practical Application

Let doing be your teacher.

Alex Podolinsky

Chapter 11

Preparation 500 Storage and Application

Preparation 500 is the foundational soil spray in Biodynamics. It fosters microbial activity, root growth and humus formation in the soil, leading to the development of friable, open soil structure. It is the most powerful substance known for the biological development of soils, but, being a living biological substance, it must be stored (with access to air), stirred and applied correctly to sustain its full effectiveness.

There are two forms of 500 used by Australian Demeter Biodynamic method growers worldwide: straight 500, which is the 500 that comes out of the cow horns dug up in spring; and prepared 500, which is the unique Australian Demeter development, made by adding the six compost

Figure 11.1. Moist, colloidal prepared 500, 0.5-acre (0.2-hectare) amount.

preparations to straight 500 around three months after lifting. Prepared 500 has proven considerably more effective than straight 500; the 500 goes through a further biological development for a time, after which it is ready to be used. Prepared 500 is the only 500 used by Australian Demeter Biodynamic growers except where the soil is initially in an extremely poor or very chemicalised state. In these situations, it is best to use straight 500 for the first few applications, since the soil can't yet take the 'full orchestra', as Alex Podolinsky characterised it. Where Biodynamic compost is regularly applied to the whole area (e.g. a market garden), straight 500 is appropriate. In practice very few operations would apply compost to the whole area each year, so prepared 500 is used in nearly all situations.

In Australia, 500 is sent by express post to minimise time in transit (unless picked up from a supplier or at the annual BDAAA conference). As soon as it is received, it should be checked for moisture and placed in the recommended storage box. Preparation 500 must always be kept in a soft, moist, colloidal state. If it begins to feel any dryer (compare with the moistness of fresh worm castings), add a few drops of pure water. Rainwater or very clean river or dam water is good, but don't use chlorinated tap water. Don't over-wet. Preparation 500 should be checked regularly for moisture. Larger quantities can also be gently turned over periodically.

Preparation 500 Storage Box

All Biodynamic growers need to have a suitable storage box for 500. Smallholders may think they don't need one, because they will spray the 500 as soon as they receive it. However, in practice, one may have to wait a week or more to find the right conditions to apply it, and 500 must be stored correctly during this time.

Where to store the 500 box

The 500 box should be kept in a coolish place, away from excessive heat and drying wind and away from electrical wiring (electromagnetic radiation adversely affects the 500), vehicles and any fumes such as petrol, oil, paint or diesel. Suitable places include sheds, shearing sheds (in or under them) and verandahs (on the south side in the southern hemisphere, to minimise sun heat; the north side in the northern hemisphere), under a house or water tank stand.

Suitable materials for the box

Natural untreated timber, preferably hardwood (some timber companies paint pesticides on kiln-dried pine). The top and bottom of the box and lid can be made from ordinary Masonite if desired. Tongue and groove hardwood floorboards provide a good seal between boards, preventing any possibility of dust from the peat moss insulation contaminating the 500. No glues or paints are to be used, nor are composite timbers (apart from Masonite) such as MDF,

plywood or chipboard, which exude chemicals. Unbleached waxed paper can be used to line the box and lid before the box is filled with peat moss; this can help prevent any peat moss dust from contaminating the 500. Also, any cracks or gaps that develop in the box or lid can be filled with softened beeswax.

Size of box

Match this to the amount of 500 you need for a year, or per season if you apply twice a year and buy fresh before each season of application. Several examples of storage boxes are shown in Figures 11.2–11.5 (overleaf). Note that a gap the width of a match is to be created between the box and the lid. Pan head screws on the top of the box will create this gap.

Insulation

Dry peat moss is tightly packed into the box and lid to act as insulation for the 500. At least 80mm of peat moss should be between the 500 and the sides and bottom of the box. Peat moss is packed in very tightly from the bottom of the box and the bottom covering is attached last. Peat moss is often moist when the bag is opened, and must be spread out to thoroughly dry before it is packed in the box and lid. Some imported peat moss originates in Europe and is conceivably contaminated by nuclear fallout from Chernobyl. It may also have been treated with chemicals before export. Avoid peat moss from those countries worst affected by Chernobyl. At the time of writing, the best source of peat moss for Australian and New Zealand growers is KiwiPeat, a New Zealand peat moss that is steam treated before export. Note that it is

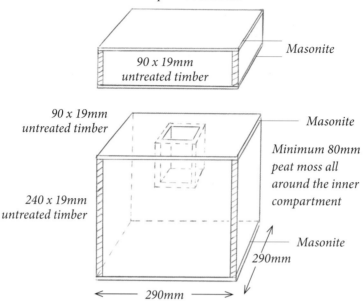

LID filled with at least 80mm thickness of dry peat moss or acceptable substitute

Masonite

90 x 19mm untreated timber

90 x 19mm untreated timber

Masonite

Minimum 80mm peat moss all around the inner compartment

240 x 19mm untreated timber

Masonite

290mm

290mm

Figure 11.2. For small amounts of 500, up to 3 acres (1.2 hectares). Small glass jar goes in inner wooden compartment. Masonite top and bottom of box and lid. Pan head screws at each corner of box top to create an air gap between box and lid. Inner compartment 50–60mm square inside (88–98mm external dimensions) and 90mm deep (inside).

Figure 11.3. For medium amounts of 500: inner wooden compartment, glazed earthenware 500 receptacle(s) with two or three drainage holes. A lid on the ceramic containers may not be needed when the containers have enough 500 in them, since moisture is then retained well. As the level goes down, or if you feel the 500 is drying too quickly, a lid (with air holes or propped up to allow air in) can be used. Finally, put small amounts in a glass jar inside the ceramic container. Pan head screws or, as seen here, two small pieces of masonite on top of the box create an air gap.

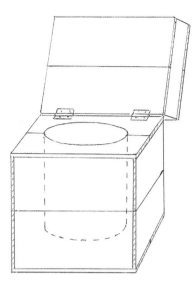

Figure 11.4. Glazed earthenware crock tightly set into the top boards of the box, or with the edges of the hole in the top boards just covering the top of the crock. However, ensure that the top boards support the crock or it will slowly sink into the peat moss, creating a dust problem. Pan head screws on top of the box create air gap between box and lid. Drainage holes in the bottom of the crock.

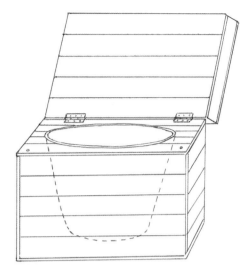

Figure 11.5. For large amounts of 500, a copper (brightly burnished with steel wool) is set into the top boards. Three 6-mm drainage holes in the bottom of the copper, covered with cow horns cut in half lengthwise. Pan head screws on top of the box create an air gap between box and lid.

peat moss not sphagnum moss that is required. Peat moss tends to reduce in volume over time, potentially leaving uninsulated air gaps inside the box and lid. Packing very tightly will reduce this problem, since any friction between loose particles gradually reduces the peat moss to dust. The bottom of the box should be screwed rather than nailed on so it can easily be removed and the peat moss insulation inspected every 4–5 years and topped up if necessary. The same applies to the lid. In future, restrictions on peat moss harvesting may lead to shortages. Investigations are underway in various countries to find acceptable substitutes, including dried peat-like materials found in natural areas on farms, vermiculite, perlite, organic or Biodynamic wool and organic or Biodynamic rice hulls. Whether any of these materials will prove to be as effective as dry, well-packed peat moss is yet to be determined and at present every effort should be made to source uncontaminated peat moss.

Preparation 500 container

Options include glazed earthenware containers (some glazes are better than others; some allow 500 to dry out), old coppers (polished inside with steel wool), undamaged enamel bowls, and glass jars (for small quantities). Apart from small glass jars, 500 containers should have some drainage holes in the bottom. Large containers require three 6-mm holes, smaller containers one or two 3-mm holes. Holes are covered with pieces of broken cow horn. When the amount of 500 remaining in the container is small in relation to the size of the receptacle, it should be put in a smaller ceramic or glass container (which can be partly covered to slow drying out) within the larger receptacle.

When and How to Apply 500

Prepared 500 should be sprayed at least once a year, preferably twice if possible. It is applied at low pressure (maximum 10 psi [70 kPa]), as large droplets, and must contact the soil, covering the whole area as evenly as possible.

The soil should be moist and warm. In many countries, spring and autumn are the most common times for application. In spring, apply after the winter chill has left the soil and before the soil dries out. In autumn, apply after the autumn break and before it gets too cold.[1] The longer the 500 has to work before winter sets in, the better. However, the autumn break is often late; sometimes 500 can be sprayed after opening rains have moistened the top few centimetres if you think follow-up rain will occur soon. Local knowledge and careful observation of weather conditions help here. Grain growers in low-rainfall areas of Australia sometimes can't sow or apply 500 until early winter or even midwinter (Australian winters are much milder than European or North American winters). Of course this is not ideal, but 500 activity still occurs, albeit less effectively than would be the case if moisture had come earlier when the soil was still warm. In subtropical and tropical climates, 500 is often best timed to coincide with the start and/or end of the wet season or during the dry season where irrigation is available.

On large acreages, particularly in low-rainfall areas, 500 is commonly applied only once per year, when suitable conditions occur, often in conjunction with sowing a crop in autumn. It is often the case that suitable conditions only occur once a year. Moreover, the logistics of getting 500 out over thousands of acres make once-a-year application a necessity.

Preparation 500 should be applied:

- After grazing pasture (and after pasture harrowing to spread the manure in the case of cattle, provided the soil is not too wet and that dung beetles are inactive).
- When grass cover is naturally low, such as at the end of summer (low-summer-rainfall areas), and when there is sufficient soil moisture.
- After slashing, topping or mulching pasture.
- On orchards, vineyards, berries, etc., after slashing or mulching grass.
- After any mulches have either mostly broken down and become thin, or have been removed from the soil (mulch can be returned after spraying).
- At or soon after sowing or transplanting a crop, or in the early growth of a crop (this also applies to a green manure crop).
- After chopping up and working in a green manure, crop residue, etc. to accelerate the transformation of the organic matter into colloidal humus.

Conditions for Application

There is no point applying 500 repeatedly on one crop; once is enough. The effect continues for the life of the crop; repeated applications on one crop can overdo it. It can be applied more often in rotational grazing situations where ongoing moisture is present in the soil; this mostly applies in irrigation dairying and can greatly accelerate soil development. Usually, at least 6 weeks should elapse between 500 applications. Preparation 500 works best where actively growing roots are present; there is little point spraying it on bare soil apart from its use on worked-in green manures.

Preparation 500 should be applied between 3pm and 2am (4pm to 3am daylight saving time). The earth's moisture belt begins to descend after 3pm, helping to draw the 500 into the soil. It begins to rise again after 3am, continuing until 3pm, making this an unsuitable period. There should be little wind, or heat, or rain (light drizzle is okay), no frost expected overnight, no heavy rain expected for at least 6 hours (this will dilute the 500 too much before it has a chance to get to work). An overcast afternoon/evening is ideal. Local weather conditions are the paramount consideration; moon and zodiac phases are much less important, though a descending moon towards new moon will be slightly more effective. Where large acreages have to be sprayed in a short window of opportunity, 500 can be sprayed all day if still, overcast conditions prevail, ideally with light drizzly rain falling.

We always aim for the best possible results from 500 applications. Following the guidelines as closely as possible will achieve this. However, there may be times when less than ideal conditions prevail, often for extended periods. Sometimes it is better to get 500 out anyway, rather than delaying

again and again and finding that conditions have become too dry, too cold, etc. Biodynamic orchardist Lynton Greenwood once remarked, 'The only place 500 does no good is in the box.'

Stirring 500 (and 501)

Preparation 500 is to be activated by stirring in pure, warm water for exactly 1 hour before applying. Stirring should be done under the open sky, not under a roof. The water is stirred as a whole so that a deep, fast-moving vortex is created, then the stirring is reversed, creating a vigorous, bubbling chaos. Another deep vortex is then formed in the opposite direction before reversing again. This process continues for 1 hour. There are several benefits from the process, the most obvious one being that the oxygen content of the warm water is greatly increased (by 60–70%), preparing the rich microbial content for maximum activity in the soil. As the oxygen breathes out again within 2 hours, the stirred 500 must be applied within 1 hour of the end of stirring for maximum benefit.

Vortex development

Whether the stirring is by hand or machine, the vortex needs to be fast moving, deep (the bottom of the vortex should be about two-thirds of the way from the top of the vortex to the bottom of the vessel) and straight sided (no round shoulders). The chaos must be vigorous and abundantly bubbling. Too slow and weak a vortex/chaos will result in insufficient oxygenation. Too strong a chaos will result in splashing and loss of water and will also roughly 'bash' the water, which is not conducive to proper activation of the 500 liquid.

Figure 11.6. A 20-acre stirring machine nearing full vortex development. Early machines had no guard on the top drive belt. This is not safe and a drive belt guard should be fitted, as in Figure 11.7. Photo: Darren Aitken.

https://videos.files.wordpress.com/UN5It9f0/vid20240415113607.mp4

https://videos.files.wordpress.com/gcmh3eVm/vid20240422095047.mp4

Water

Use the purest water you can find. This may be rainwater or pure spring, stream, river, bore or dam water. Town water is not suitable, unless stored outside in an open stainless steel or copper vessel for at least a week and stirred every day for a few minutes. Even then, it is not recommended.

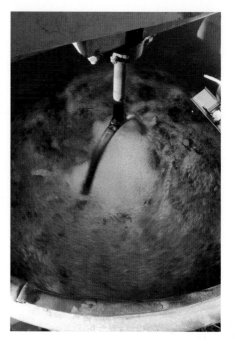

Figure 11.7. An 8-acre machine – with chaos. In this design, a stainless steel swivel bar is suspended from a slightly bendable arm (to allow adjustment). When the vortex reaches optimum development the water pushes the bar to left or right, switching off the motor. As soon as the water level drops enough, the bar returns to its horizontal position, switching the motor on again, in the reverse direction.

Warming the water

For 20-acre stirring machines, the water should be warmed to between 28°C and 33°C. Never exceed 35°C. The ideal is for the water to be heated evenly, as a whole, rather than, for instance, adding very hot water to cold water, but the latter still works well. Water may be heated with a wood, gas or diesel burner (don't allow fumes to waft over the stirring machine) or the sun (in subtropical/tropical areas, but be careful that the temperature doesn't exceed 35°C). Electricity

Figure 11.8. Heating water for 500: a gas ring under the tank warms the water before gravity feeds it into the stirring machine. A new batch can be warming while the operator is spraying out stirred 500. This is only one idea. Many other heating solutions can be found.

has a deleterious effect on the water, making it less effective for 500 activation. To stir smaller amounts of water, from hand stirring up to 8-acre stirring machines, heat the water as a whole to 35°C, no higher. Small quantities of water can be a little warmer initially than larger quantities, to help heat to be retained longer. Larger quantities of water retain heat much better.

Flowforms

Some Biodynamic practitioners advocate the use of 'flowforms' to 'stir' 500. Flowforms are a series of bowls through which water flows similarly to natural streams, and no doubt do add oxygen to the water. However, the water spends most of the time sitting in a holding tank waiting to be pumped to the top of the flowform and is not activated as a whole. Far less oxygen is incorporated into the water than with

a single deep, fast-moving vortex alternating with a vigorous bubbling chaos. Also, picture the 'drawing power' of a single deep vortex, akin to the action of the sun on the planets as described by Podolinsky.

Stirring vessels

The water should be stirred in a vessel made of copper, stainless steel, undamaged enamel or glazed earthenware. A concave base is ideal if possible, but is not essential. The vessel should never be used for any purpose other than stirring 500 or 501 and should be covered when not in use, to keep it clean.

Quantities

Preparation 500 (including prepared 500) is applied at the rate of 35g of 500 stirred in a minimum of 3 imperial gallons (13.6 litres) of water per acre (88g in a minimum of 33.7 litres per hectare).[2] The 500 is broken up in the water just before stirring starts, or added after starting the machine, the timing starting from when the 500 is added. The essential measure is the amount of 500 per acre; this can be stirred in up to twice the normal proportionate volume of water, but no more, or it becomes too dilute and less effective (akin to heavy rain occurring within 6 hours of spraying). The amount of water in a stirring machine should never be less than what it is designed for, e.g. 24 imperial gallons (109 litres) in an 8-acre machine, 60 imperial gallons (273 litres) in a 20-acre machine, but the amount of 500 can be lowered. The reasons for stirring a lower than normal amount of 500 in the set volume of water include the following:

- For small gardens, the minimum amount of water in which a satisfactory vortex and chaos can be created is 1.5 imperial gallons (6.8 litres) or half the normal amount for 1 acre (0.4 hectare) of 500. Depending on the size of garden, from 0.25 acre up to 0.5 acre of 500 can be stirred in this amount of water. For smaller gardens, apply the correct amount of stirred 500 so that the appropriate amount of 500 will be used for the area. For example, for a 0.1 acre (400m²) garden, stir 9 g of 500 in 6.8 litres of water and apply 2.7 litres (6.8 × 400 ÷ 1,000) to the area. (Pour the rest on your compost heap.)

- To cater for areas that don't exactly match the size of your stirring machine. For instance, a 5 acre (2 hectare) paddock is to be sprayed and you have an 8-acre machine. Weigh out 5 acres (2 hectares) of 500 and stir in 24 imperial gallons (109 litres) of water. Obviously, you will have to spray this out at a slower speed than when you have stirred 8 acres (3.2 hectares) of 500 in the machine. Record tractor gear and rpm for each area sprayed (4–8 acres [1.6–3.2 hectares] for this size machine). You will soon learn to drive at the correct speed for each area sprayed.

- Where grass is slashed rather than grazed before spraying. This may include (but is not limited to) orchards, vineyards, *Rubus* berry plantings, and olive groves. In these cases it can be advantageous to spray up to twice the amount of water per amount of 500 to ensure better contact of the liquid with the soil. The cut grass reduces contact with the soil.

Figure 11.9. Hand stirring vessels – earthenware, stainless steel, copper.

Figure 11.10. Hand stirring vessels, 0.25–2 acres (0.1–0.8 hectare). Photo: Keith McCallum.

Apart from the above three situations, it is best to use the normal recommended amount of 500 per volume of water (35 g of 500 per 3 imperial gallons [13.6 litres] of water per acre [88g per 33.7 litres per hectare]).

Stirring machines

Since the 1960s, Australian-designed stirring machines that replicate the best qualities of hand stirring have been used to enable Biodynamics to be efficiently applied on broad acreages. An important design feature is that the vortex/chaos is not timed, but is directed solely by the vortex development; the paddles do not reverse until a good deep vortex has formed. These machines have proven more effective than hand stirring for any amount of water over 12 imperial gallons (54 litres). Machines can be made in various sizes; the most common in Australia being 24 imperial gallons (109 litres) (for up to 8 acres [3.2 hectares]) and 60 imperial gallons (273 litres) (for up to 20 acres [8 hectares]). The 20-acre machine is the largest recommended for ideal vortex/chaos development. For large properties,

Figure 11.12. Stirring machines made by BioMeccanica, Italy. Photo: Saverio Petrilli.

multiple 20-acre barrels (up to six or even more for very large properties), with a single drive mechanism so that only one motor and one set of switching apparatus are required to operate all the connected machines. Some farmers make their own stirring machines (and spray rigs) and BDAAA or BDGAI can give contact details for machine-makers. Eco-dyn in France (www.ecodyn.fr) and BioMeccanica, in Italy, also make stirring and spray-out equipment and have agents in North America (see 'Biodynamic Resources' section). The machine makers can advise on the best-size set-up for your particular property, as can experienced Biodynamic farmers.

Smaller, 4-acre electric stirring machines are also available, through BDGAI, as are small machines operated by manually turning a handle that drives (reversible) paddles, in 1-, 2- and 3-acre sizes. These small machines take a lot of the work out of hand stirring and are very good for smallholdings.

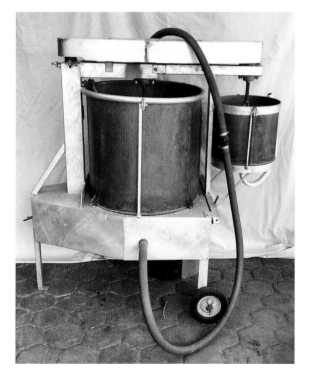

Figure 11.11. An 8-acre stirring machine with a 1-acre barrel on the side, both driven by the same belt. Note the belt guard at the top.

Adjustment

Stirring machines periodically need adjustment to create a good vortex and chaos and to equalise

Figure 11.13. Four-barrel stirring set-up. Machines operate in tandem with a single switching mechanism that reverses the action simultaneously in all barrels, and on a high platform so that the stirred 500 can be drained quickly into the spray tank.

Figure 11.14. Four-barrel stirring set-up.

Figure 11.15. Five-barrel set-up, Burrum Biodynamics, Marnoo, Victoria.

the time spent running in each direction. Some machines use a float mechanism to switch off the motor when optimum vortex development is achieved (and switch it back on as the water level at the side of the barrel falls). The height at which the float trips the switch can be adjusted to create a deeper or shallower vortex. The vortex should extend about two-thirds of the way down from the top of the vortex to the bottom of the vessel. Other machines use a pivoting mechanism in the water which switches the motor off when the vortex is high enough to flip it, and similarly switches it back on as the water level falls. This pivoting switch can be carefully bent from side to side to equalise the time for clockwise and anticlockwise vortex development, and bent up or down to make the vortex deeper or shallower. Some of the pivot mechanism machines develop a slight 'back-

wave' when chaos ensues, briefly cutting and restarting the motor. This disturbs the process as well as putting extra stress on the motor. If it occurs, try slightly reducing the water level (say, by 1cm) and adding a small weight such as a short length of light chain to the switch outside the barrel. The drive belt of machines can be tightened to create a faster, stronger vortex and chaos or loosened to moderate the action. Tightening the belt too much can create

Figure 11.16. Mobile six-bowl stirring set-up, Cashmores, Western Australia. 120 acres (48 hectares) per stir. Photo: John and Bernadette Cashmore.

Figure 11.17. Mobile six-bowl stirring set-up. Photo: John and Bernadette Cashmore.

excessive strain on the electric motor in the case of electrically driven machines. Some machines need the drive belt to be adjusted over the first 5 minutes, until it has warmed up. Initially it needs loosening to reduce the vigour of the chaos; later it needs tightening as it warms up and the chaos becomes too soft.

Hand Stirring

Whether your property is large or small, it is advisable to learn how to stir by hand. It helps when evaluating the operation of your stirring

machine and making any necessary adjustments, and is also necessary if you make Biodynamic compost, since the valerian compost preparation must be hand stirred for 20 minutes before it is added to the compost heap.

There are some critical points to remember when hand stirring. Begin in the middle, with a limp wrist and a fast stirring action. This gets the water moving fast from the start (if it is too slow, it will be impossible to create a deep, straight-sided vortex later). Once the water is moving fast, pivot more from the elbow and, with a firmer wrist, move the stirring action outwards. This deepens the vortex and straightens the sides. As soon as a deep, straight-sided (not round-shouldered) vortex is formed, reverse the direction of stirring by moving back into the centre and starting again as above. The chaos should be vigorous and bubbling without, however, being too rough and bashing the water (resulting in splashing). There are considerable variations between people in their individual hand size and stirring style. This is why, on small properties or on gardens, only one person should do the stirring for the required hour.

It is important to relax into the task, and not be too stiff, since you will tire easily. When one arm tires, change arms by removing the stirring hand from the water as soon as a deep vortex has been formed and replace with the other hand, stirring in the opposite direction to create the chaos. It is usually easier to make the change after a forward-stirred vortex: for example, create the vortex with the left hand, stirring clockwise, then withdraw the left hand and create the chaos with the right hand, stirring anticlockwise. In your early efforts at stirring, there will be some difference between the two arms, but with practice this will become minimal.[3]

Spraying 500

As soon as possible after the 1 hour of stirring time finishes, the 500 liquid should be sprayed out. The oxygen dissolved in the water gradually breathes out, slowly reducing the effectiveness of the 500. Therefore, the 500 should be applied within 1 hour of the end of the stirring time. It is sprayed so that the droplets contact the soil. On smaller properties, 500 can be sprayed with a stainless steel, copper or brass knapsack (or even flicked out with a natural bristle brush or a leafy branch) that has never been used for any purpose other than spraying 500 or 501, but on larger properties a spray rig (same materials and also only ever used for spraying 500 or 501) matched to the size of the stirring machine is necessary. Plastic (or fibreglass) knapsacks or spray tanks are not suitable because of the leaching of chemicals and tiny particles, which reduce the effectiveness of the 500. Because the 500 is only briefly in contact with pipes and nozzles as it is pumped out (and for reasons of practicality), these may be plastic.

Preparation 500 is to be sprayed at low pressure, no more than around 10 psi (70 kPa), creating large droplets that evenly cover the ground. The pump should be of a diaphragm or roller type, not a centrifugal type.

There are two main types of spray rig for 500: the boom spray, in which flat fan nozzles are mounted on a boom, evenly covering the area sprayed, and a reciprocating unit, in which

Figure 11.18. Spraying 500 in a vineyard. Photo: Rob and Pauline Bryans.

Figure 11.19. An 8-acre 500 spray rig with reciprocating action. Covers a width of 10 m with large droplets.

two round nozzles swing side to side, driven by a windscreen wiper motor, emitting large droplets, evenly covering a 10–20-m width, depending on the set-up. Spray units can be three-point linkage for use with tractors (and can also be mounted on the tray of a ute), or trailed units for pulling

Figure 11.20. 300 imperial gallon (1,400 litres) spray tank – 100 acres (40 hectares) per application.

Figure 11.21. Cashmore's 500 spray rig: 120 acres (48 hectares) (360 imperial gallons [1,600 litres]), 37-m coverage boom spray, 120 acres (48 hectares) per hour.

behind quad bikes. For large acreages, spray rigs have been built that cover up to 37-m width, allowing 500 to be sprayed at the rate of 120 acres (48 hectares) per hour.

Sieving the Liquid

Stirred 500 must be sieved as it moves from the stirring machine to the spray tank, to remove

fibres that could block the spray nozzles. A fine sieve is used, whose holes should be no more than half the size of the nozzle apertures.

Covering the Ground

When spraying with a knapsack, covering the required area without running out or having too much left over at the end is a matter of experience. It is advisable for beginners to do a few test runs with water in the knapsack, to learn how fast to walk and how to distribute the liquid evenly. If spraying at the normal rate of 35g per 3 imperial gallons per acre (88g per 33.7 litres per hectare), it is necessary to wave the spray wand side to side quickly as you walk, keeping a close eye on where you have been. With market gardens, the beds or rows act as a useful guide. If you have stirred half the normal amount of 500 (e.g. 9g of 500 in 1.5 imperial gallons [6.8 litres] of water for a 1000m² area), a flat fan nozzle can be used and you can walk fairly quickly keeping the spray wand pointed straight ahead and not swinging it back and forth.

When spraying on a larger scale, we choose the lightest practicable vehicle: for smaller areas, a quad bike pulling a trailed spray tank is ideal. For larger areas the lighter the tractor the better. A ute is often used with the tank mounted in the tray. However, when applying 500 to broadacre farms, larger tractors may be required. Some farmers use vehicles with balloon tyres to spread the weight more evenly over the soil surface. As Podolinsky cautioned, however, the reduction of the weight applied per square inch by use of balloon tyres or caterpillar tracks cannot prevent the overall weight of the vehicle pressing on the subsoil. He posed the question of whether you

Figure 11.22. Cashmore's – filling spray tank with prepared 500 after stirring.
Photo: John and Bernadette Cashmore.

Figure 11.23. Cashmore's spray rig at work – 120 acres (48 hectares) of 500 per batch.
Photo: John and Bernadette Cashmore.

would want to drive a vehicle weighing 2 tonnes over a bridge with a 1-tonne carrying capacity.

Various methods are used when spraying by tractor:

- The most accurate is to use a GPS. Initially follow a fence line in such a way that the spray droplets just reach the fence. When you reach the end of the field, turn around (keep the pump running during the turn) so that the edge of your spray pattern just reaches the edge of the previous spray and set the GPS at 180 degrees to the original direction.
- Foam markers set at the outward ends of a boom sprayer can be used as a guide, bearing in mind that the spray pattern may, depending on the set-up, extend further than the ends of the boom and must be allowed for.
- On smaller properties, bags can be placed over steel posts at the ends of the field before stirring, set at the right distance apart to suit your spray width. These can be used as a sight guide. Similarly, different-coloured reflectors can be attached to fences at the ends of fields to act as a guide when spraying after dark.

- Spray without any guides, just using your judgement as to where you sprayed last. Features such as trees, hills or harrowing marks can assist. Some overlap is not a problem, though ideally you should cover the ground as evenly as possible.

After Spraying

If another batch of 500 was stirring while you were applying the first, refill and keep going. Otherwise, rinse out the stirring machine with clean water, then drain and cover it to keep it clean. It is also a good idea to pump some clean water through the spray rig or knapsack and drain this before storage.

https://videos.files.wordpress.com/bfO6wPPu/
vid20240626070158.mp4

Table 11.1. Preparation 500. Spray Quantities Guide (see note overleaf)

Area to be sprayed	Stirring vessel	Volume of water (litres)	Amount of 500 (gm)	Amount to spray (litres)
100m²	20 litre s/s bucket with or without hand stirring frame	6.8	9	0.7
200m²	"	6.8	9	1.4
500m²	"	6.8	9	3.4
0.25 acre (1000 m²)	"	6.8	9	All
1500m²	"	6.8	14	All
0.5 acre (2000 m²)	"	6.8	18	All
1 acre (4000 m²)	Copper, large stainless steel vessel, hand stirring machine	13.6 (3 imperial gallons)	35	All
2 acres (0.8 hectares)	"	27.3 (6 imperial gallons)	70	All
3 acres (1.2 hectares)	"	41 litres (9 gallons)	105	All
4 acres (1.6 hectares)	8-acre stirring machine	109 (24 imperial gallons)	140	All
5 acres (2 hectares)	"	109	175	All
6 acres (2.4 hectares)	"	109	210	All
7 acres (2.8 hectares)	"	109	245	All
8 acres (3.2 hectares)	"	109	280	All
10 acres (4 hectares)	20-acre stirring machine	273 (60 imperial gallons)	350	All
11 acres (4.4 hectares)	"	273	385	All
12 acres (4.8 hectares)	"	273	420	All
13 acres (5.2 hectares)	"	273	455	All
14 acres (5.6 hectares)	"	273	490	All
15 acres (6 hectares)	"	273	525	All
16 acres (6.4 hectares)	"	273	560	All
17 acres (6.8 hectares)	"	273	595	All
18 acres (7.2 hectares)	"	273	630	All
19 acres (7.6 hectares)	"	273	665	All
20 acres (8 hectares)	"	273	700	All

Spraying 500 and 501 Together

In emergency situations, 500 and 501 can be sprayed together, after stirring for 1 hour, to quickly strengthen plants that are under attack. This has been used most successfully, for instance, against red-legged earth mite and lucerne flea in crops and pastures that are under attack (see pp. 87–88).

Notes for Table 11.1

For other areas under 0.25 acre (1000 m²), use the formula:

Litres to spray = 6.8 × area in m² ÷ 1000

Stirring machines have a mark or depth gauge for the correct volume of water.

The amount of 500 was originally determined by Dr Ehrenfried Pfeiffer's research and relates only to soft, moist, fully colloidal 500. Some Biodynamic sources quote larger amounts per acre, no doubt because the material was dried after removal from the cow horns. Once this substance is dried, practitioners of the Australian Demeter Biodynamic method regard it as no longer 500 and would expect it to have little effect compared with fresh, moist, high-quality colloidal 500.

FOOTNOTES FOR CHAPTER 11

1. In Australia the 'autumn break' is when consistent rain comes after the dryness of summer.

2. These quantities were determined by Ehrenfried Pfeiffer's research and apply to moist, fully colloidal 500.

3. See John Bradshaw, 'Stirring and Applying 500 and 501', *Biodynamic Growing*, 2 (2004), pp. 3–5.

Chapter 12

Applying 501 – the Light Spray

The Biodynamic preparation 501 is made by adding water to very finely ground, clear quartz crystals to make a paste, and burying this in cow horns from spring until autumn. It is the partner of 500. Preparation 500 dynamically activates soil biology, building humus levels and microbial activity and stimulating root growth. Preparation 501 stimulates upper plant activity, greatly enhancing light metabolism, bringing a crispness and crystallinity to plants, balancing over-lushness and tendencies towards disease. It is a most powerful Biodynamic tool and, used when conditions warrant, will greatly enhance the flavour, sweetness and keeping quality of crops. Alex Podolinsky used 501 on his cherries during wet springs in the 1950s to keep them healthy when there were no other cherries in the Melbourne market because of widespread disease.

Figure 12.1. Clear quartz crystals suitable for making 501.

When to Use 501

Use of 501 can commence after 500 has begun the soil-structuring process. It is used to bring more light into plants when they are exposed to conditions that predispose them towards over-lushness – something that may be caused by low light conditions, excessively watery conditions, over-rich soil or a combination of these factors.

Examples include:

- Extended periods of overcast, cloudy weather.
- Ongoing excessive rain or humidity.
- Over-watering.
- Warm conditions combined with ongoing cloudy weather: plant feeding is stimulated by sun warmth, but lack of light reduces photosynthetic activity, affecting the plant's ability to fully assimilate nutrients.
- When plants are 'pushed' by excessively rich soil.

Any of these conditions, resulting in over-lushness, make plants more susceptible to attack by disease or insect pests, as well as causing the produce to have poorer flavour. Podolinsky called such plants 'overblown'. An experienced observer, used to assessing upper plant expression, can immediately see this and take action.

Preparation 501 can be applied whenever such conditions apply. The effect will last for a month or more, but under extremely challenging conditions 501 can be applied more often. It is most commonly used in spring and summer, but autumn and winter applications can bring exquisite flavour to winter vegetables.

Figure 12.2. Preparation 501.

When it is sprayed on autumn-saved pasture, the grass will retain excellent quality into winter.

Apply 501 in the following circumstances:

- Early in a plant's growth, but avoid spraying young seedlings. If the area to be sprayed includes young seedlings, they can be temporarily covered until spraying has finished.
- On transplanted seedlings, wait until they have settled in and new growth is underway. However, seedlings that are well established but leggy can be sprayed a week or so before transplanting. Avoid spraying 501 on lettuces before they are transplanted and well settled in.
- On fruit trees, apply when the leaves are well established, in early to mid spring and when the fruit is set but still very small. Apply again when the fruit is fully grown, to aid ripening and enhance flavour.

- On grape vines, when the leaves show five points, before flowering commences. In wet seasons, when fungal diseases threaten, spray again when the fruit is set, until it ripens. Preparation 501 can also be sprayed 1–3 weeks before picking to further enhance the quality of the fruit or, if weather conditions threaten, to reduce the chance of fungal diseases.
- On pasture, when the fresh growth is around 100mm high, grazing around 10 days later. In wet springs when cattle are scouring, 501 can crispen the lush grass and dry up the manure. But be judicious – it can also reduce the amount of feed if conditions are drying.

Cautions with 501

Preparation 501 must be used carefully and only when conditions warrant:

- Preparation 501 increases transpiration, so extra water may be advisable for a few days after spraying if conditions are drying.
- Do not apply after a dry spell where irrigation has not been used. It may bring too much light, drying the plant up or pushing it to seed prematurely.
- Avoid spraying vegetables or fruit trees that are flowering, because the flowers will dry up. Exceptions are cucurbits (whose flowers are watery), and plants that flower progressively such as tomatoes and strawberries.
- In large parts of Australia, particularly grain-growing areas, rainfall is low and sunlight is very intense (far stronger, for instance, than in Europe). Generally, 501

is not needed in these areas, since there is already an abundance of sunlight. However, over winter, long periods of cloudy and sometimes rainy weather can occur. Preparation 501 can be beneficial here too. However, less 501 should be used per volume of water and per hectare. Careful experimentation should be done to determine the best rate for each area, applying the 501 to a small area of the crop one year (even several applications to different small areas at different strengths) to observe the effect, before applying it more generally to subsequent crops. There is the danger that too much 501 can damage a grain crop. One of the problems with applying 501 to a grain or legume crop when overly wet periods prevail is that the fields are too wet to drive on. Drone application may be possible in the future.

- In subtropical and tropical areas, sunlight is very intense. Preparation 501 is generally not needed at all during the dry season. It is more relevant during the wet season, but caution is needed here too, since sunlight at this time can still be very intense. Experience is the best guide. Spraying in the late afternoon can be tried, to bring the beneficial light influence without so much risk of damage.

Storage

Preparation 501 is stored in a glass jar with the lid loosened half a turn so some air can circulate. Place the jar on a windowsill that receives good light, including direct morning sunlight, that is, an east-facing window. Store away from any electrical wiring or fumes. Turn the jar periodically and invert it every few months to agitate the 501 before turning it the right way up. Tighten the lid before inverting and loosen it again when the jar is upright.

Stirring

Preparation 501 is stirred like 500 (see Chapter 11), in pure water heated as for 500, for 1 hour (see p. 72). Only 1g per 3 imperial gallons (13.6 litres) of water is needed per acre (2.5g per 33.7 litres of water per hectare). For 0.5 acre (2000m^2), use 0.5g of 501 in 6.8 litres of water. For 0.25 acre (1000m^2), use 0.25g of 501 stirred in 6.8 litres of water and spray the whole amount. For areas less than 0.25 acre (1000 m2), stir as for 0.25 acre (1000m^2) but spray an amount of liquid proportionate to the area: litres to spray = 6.8 × area in square metres ÷ 1,000.

Quantities to use may vary somewhat from country to country, and farmers should consult their own national Biodynamic association. In Australia, the quartz crystals are crushed into small salt-like particles (and any metal particles left after pounding removed with magnets) before being dry ground to talcum powder fineness between two sheets of glass in a purpose-built machine. This, after burial in cow horns from spring till autumn, results in highly effective 501. European users of Australian 501 report that it is significantly more effective than 501 made from quartz crystals that have been ground in water, possibly because the dry grinding retains the crystalline structure of the fine powdery particles better than does water grinding, which may tend to round off the crystals somewhat.

Spraying 501

As with 500, 501 should be sprayed as soon as possible after stirring ends, finishing within 1 hour. It is generally sprayed on a bright sunny morning, finishing before 10am, earlier in summer. If a prolonged cloudy/humid/rainy period persists and a sunny morning can't be found, it can be tried in less ideal conditions, giving a reduced though still beneficial effect. The morning must be as still as possible: 501 is applied as a fine mist with high pressure up in the air so that it slowly drifts down on to the foliage. If your spray set-up pumps the 501 directly through a small nozzle or series of nozzles, use at least 40 psi (275 kPa), up to 80 psi (550 kPa). If you are using a pressured stream of air to turn pumped droplets

Figure 12.3. Preparation 501 spray set-up for a knapsack, Bourgogne, France.

of 501 into a fine mist, less pressure can be used to pump the 501. For small areas, 501 can't be applied without a knapsack. Use a stainless steel, copper or brass knapsack, attaching a smaller nozzle (or pair of nozzles) than for 500; the same knapsack can be used as you use for 500. Don't use the knapsack for any other purpose.

Any light breeze will cause excessive drift. Even on a seemingly completely still morning, there will be some drift. Observe the drift and adjust accordingly the way you move over the area. For instance, it is no use starting exactly at the edge of a vegetable crop if the spray doesn't land on the first few rows; you would need to start further away so that the spray actually lands on the first rows.

Figure 12.4. Hand spraying 501 on strawberries, Turingal Farm, New South Wales.
Photo: Graeme Gerrard.

Figure 12.5. Spraying 501 in a vineyard, Tuscany, Italy. Photo: Saverio Petrilli.

A 501 Trial

Saverio Petrilli, vigneron and Australian Biodynamic method adviser in Italy, reported on a 2009 trial at the vineyard he manages, Tenuta di Valgiano, in Tuscany.[1] The year 2009 was particularly difficult, with heavy rain for 3 months in winter, followed by an unusually hot and dry period in summer. June in the area can be very humid because the clay soils slowly release humidity under the hot summer sun. One section of vineyard received three sprays of 501 during spring, while another section received no 501. The section sprayed with 501 performed far better in terms of its ability to withstand both powdery mildew and, surprisingly, heat damage, as can be seen in Figures 12.6 and 12.7 (overleaf).

Spraying 500 and 501 Together

In emergency situations, 500 and 501 can be sprayed together (after stirring for 1 hour) to

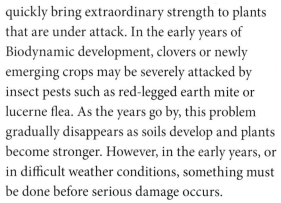

Figure 12.6. Vermentino grapes 2009 without 501. Photo: Saverio Petrilli.

Figure 12.7. Vermentino grapes 2009 with 501. Photo: Saverio Petrilli.

quickly bring extraordinary strength to plants that are under attack. In the early years of Biodynamic development, clovers or newly emerging crops may be severely attacked by insect pests such as red-legged earth mite or lucerne flea. As the years go by, this problem gradually disappears as soils develop and plants become stronger. However, in the early years, or in difficult weather conditions, something must be done before serious damage occurs.

Spraying 500 and 501 together in emergency situations was first suggested by Podolinsky and has been proven effective by many Biodynamic farmers. Farmers report that, after applying 500 and 501 together, red-legged earth mites can be seen moving around but not staying on the plants, as if the plants are no longer palatable, and cause no further damage. Preparations 500 and 501 together are sprayed at low pressure, with more effect on the upper plant if sprayed in the early morning, and more effect on the roots if sprayed in the mid to late afternoon. The mixture can, however, be sprayed at any time of day. An alternative is to spray 500 one afternoon and 501 the next

morning. This can possibly have an even better effect than spraying the two together. However, it requires twice the time and resources, and weather conditions the morning after the 500 spray may well not be suitable for spraying 501.

FOOTNOTE FOR CHAPTER 12

1.   See Saverio Petrilli, 'Two Vineyards on the Hills of Lucca', *Biodynamic Growing*, 14 (2010), pp. 15–20.

Chapter 13

Biodynamic Compost-Making

Nature, in its infinite wisdom, is intricately organised towards the feeding of plants through soil colloidal humus. The myriad living organisms in and around the soil work endlessly on anything that has once lived, to break it down and then build it into highest-quality colloidal humus, the perfect natural plant food.

These living organisms range from the tiniest micro-fauna and micro-flora, including microbes and fungi, assisted by the enzymes and co-enzymes produced by animals, plants, moulds, fungi, yeasts and bacteria, through the meso-fauna, such as mites, springtails and nematodes, to the macro-fauna, including ants, earthworms, beetles and termites.

As Biodynamic growers, we learn how to skilfully assist the process of turning organic matter into colloidal humus, whether it be through the application of prepared 500 to soils, in the growing of green manure crops (using prepared 500 in the growing and working in of these crops) or in the making of highest-quality, colloidal humus Biodynamic compost, using the six Biodynamic compost preparations to guide and facilitate the humus formation process.

Every Biodynamic farmer, smallholder and gardener should learn the art and science of making high-quality colloidal humus compost. Although the making and use of compost is not necessary (or practical) in many farm situations, every farm should have a home garden to supply fresh Biodynamic vegetables and fruit to the family. Every household generates vegetable scraps, weeds, grass clippings, etc. and what better than to know how to turn this organic matter into colloidal compost and use it to grow healthy vegetables and fruit for the family?

Large-scale compost-making is not generally part of beef, sheep or grain farming. However, it is a natural part of dairy farming: manure is collected from the holding yard or barn after every milking. Compost can also be very beneficial in market gardening, though many market gardeners rely solely on humus developed via green manures. Normally, green manure crops provide enough colloidal humus for two crops. Adding compost after the second crop allows the growing of a third crop on an area in one year. Market gardens that produce their own seedlings also need compost as an essential ingredient in seed-raising mixes. Handmade heaps provide enough compost for this purpose on small-scale market gardens, and usually enough extra to feed some crops.

Compost can also be very valuable in the growing of fruit trees, berries and grapes. Grape marc provides a good base material for compost-making, just needing manure, and perhaps some hay or straw to make a good blend.

It is not particularly difficult to make high-quality colloidal compost once the technique has been learnt. However, many composts made today are of very poor quality. Whether it be the compost site, the materials chosen, the method of building the heap, the moisture content or some other factor, much 'compost' produced today is either too dry, crumbly or smelly, anything but colloidal.

I made my first compost heap aged 14, guided by a well-known Australian garden guide that directed me to build the heap by making 100-mm alternating layers of green material and soil. I was most disappointed when, after months, there appeared to have been no change at all in the materials. Learning to make Biodynamic compost resulted in success every time.

Are the Biodynamic Compost Preparations Necessary?

The Biodynamic compost preparations (502–507) guide and accelerate the composting process from the start, conserving the elements in the raw materials and producing a product that is colloidal, microbially rich, well balanced and high in nitrate nitrogen. It has been suggested that each preparation works to facilitate the activity of a different element or group of elements, or to strengthen plants in different ways, and that these qualities continue to work in the soil and plants long after the application of the compost. Ray Quigley, a Victorian Biodynamic dairy farmer, applied Biodynamic compost to a field once only, before selling that field to a neighbour. He could still see the beneficial effects of that one compost application in the pasture 20 years later.

Well-made Biodynamic compost is 100% rich, dark, moist colloidal humus. It is not dry or crumbly, but can be kneaded in the hand like putty and doesn't appreciably stain the hands. A skilled practitioner *can* make high-quality colloidal humus compost without using the six Biodynamic compost preparations. But is it as good?

Tests run by Ehrenfried Pfeiffer showed that the use of the Biodynamic compost preparations produced a higher-quality compost. He ran a series of controlled trials of compost-making under laboratory conditions, in which carefully weighed amounts of fresh cow manure, straw and earth were mixed, placed loosely in glass vessels and monitored. Only sterile air was admitted to the glass vessels. One set of vessels had no Biodynamic preparations added, while the other set had 0.2g each of Biodynamic compost preparations 502–507 added.

After 10 days the materials in the Biodynamically treated vessels had already changed in colour and the manure odour had disappeared, whereas the untreated materials showed little change. After 2 months, the materials were carefully analysed. There were insignificant differences in the content of potassium, calcium, magnesium, ammonia nitrogen (trace only in both), and manganese. The Biodynamic compost had a 5% higher level of phosphate. However, the really startling finding was that the Biodynamic compost had a 233% higher level of nitrate nitrogen than the untreated compost: 61.6 pounds per ton compared with 18.48 pounds per ton. In addition, the bacteria count was vastly superior in the Biodynamically prepared compost. After 2 months the non-treated compost had 800 million bacteria per gram of material, compared with 3,000 million per gram of the Biodynamically treated compost.

The Compost Process

Much 'compost-making' these days involves excessively high temperatures: nitrogen is steamed out of the materials in the form of foul-smelling ammonia, and the valuable organic matter is essentially burnt up. The resulting 'compost' is fibrous rather than having the characteristic putty-like colloidiness. The nutrients that are left will leach out readily and feed plants indiscriminately. Other methods, particularly where no animal manures are used, result in a cold composting process and rarely end well.

Compost-making is both an art and a science. One can quickly learn the science – the essential requirements of good compost-making established over many years. The sensing side takes longer to develop. Determining the correct proportion of materials and moisture levels and the monitoring of progress are skills that develop over time.

Biodynamic compost-making is an aerobic process – that is, the materials have access to air and are not sealed up and anaerobic as in some plastic compost bins. The heap is free-standing, not enclosed in any way, and will not need turning. Once the heap is made, it does not smell and does not seem to attract or breed flies. Biodynamic compost heaps can therefore be made even on small suburban blocks without disturbing neighbours.

The composting process involves several different stages. The heap heats up to between 45°C and 50°C or, at the very most, 55°C within a week or two and may hold this temperature for several weeks before gradually declining. Heating over 55°C can result in considerable loss of nitrogen to the atmosphere. As the temperature slowly declines, active fungal growth and microbial activity occur, then micro- and macro-fauna become involved, and finally the heap is thoroughly worked by compost worms. These worms do not normally need to be added, they come when the situation is right.

Site Selection

The ideal site is sheltered from strong winds, has fertile well-drained soil, is easily accessible for delivery of materials and is not too far from where the compost will be used. Some shade in summer is desirable, but avoid the close proximity of large trees (especially eucalypts) whose roots will invade the compost. The site should not be too steep. Compost heaps should preferably run north–south to allow the sun equal access to both sides.

Shelter

If possible, plant silver birch, hazel and elder around the compost yard. These plants provide shelter and support fermentation processes on a compost site. Evergreen shrubs and trees (not eucalypts) planted further back will provide extra shelter in winter. Screens can be built to provide shelter immediately.

Fertility

Poor soil, particularly heavy clay soil, will hamper the composting process. When the finished compost is removed, leave 2–3cm of compost on the ground to assist the next heap. Cover with mulch material and keep it moist. After several years of composting on a site, even a pure clay base will improve dramatically and the compost will mature much faster.

Drainage

Poor drainage can double the time it takes to produce good compost. Poorly drained sites can be improved by digging open drains around the sides and leading water away from the site. It may also be worth installing agricultural drainpipes under the site.

Tree roots

If your only possible site is close to greedy trees, dig a trench at least 30cm deep between the trees and the heap, to cut as many roots as possible. The soil could be replaced, but the operation would have to be repeated annually. In any case, you may have to dig a bit deeper each year. Another solution is to only make compost in autumn, so that it will mature through the winter when tree roots are less active and, in the case of deciduous trees, dormant.

Compost Materials

I tire of hearing popular TV garden show presenters advising that compost materials should be a blend of 'green' (kitchen scraps, fresh leaves, weeds, grass clippings, etc.) and 'brown' (dry twigs, dead leaves, shredded paper, dried lawn clippings, straw, etc.). This is very reminiscent of my first disastrous effort at compost-making. Rarely if ever is there any mention of animal manure, one of the most important ingredients. There should be a good mixture of manures and plant materials.

Almost anything that has once lived can go into a compost heap. Possible materials include kitchen scraps, weeds (wilted, not fresh), fresh or spoilt hay, straw, leaves, lawn clippings (not fresh), sawdust (not treated pine), and manures (cow, horse, sheep, goat, hen, rabbit, pigeon, pig, etc.). This is only a small sample of possible ingredients. The two best materials are hay and cow manure.

Ideally, all materials used would be Biodynamically or organically produced. However, few people are lucky enough to have access to such materials. Compromises must be made, but we must be careful to avoid materials that are likely to be contaminated with agricultural chemicals. Conventionally grown straw may carry pesticide/fungicide/herbicide residues, and seaweeds may be contaminated with heavy metals. If in doubt, a chemical residue test can be conducted. This will likely be a requirement if you're applying for certification. Avoid manure from animals that have recently been chemically wormed. I would avoid the use of paper products in the heap, with the exception of unwaxed brown cardboard with minimal printing.

Materials should be chosen that are a mixture of high-nitrogen materials such as manures and high-carbon materials such as hay, straw, and dead leaves. The ideal carbon–nitrogen ratio for compost–making is in the range 25:1 to 30:1. Much less than 25:1 can result in overheating of the heap, leading to loss of nitrogen as ammonia. If the ratio is much higher than 30:1 the heap will not heat up enough. However, I don't recommend worrying too much about this. Not only is the carbon–nitrogen ratio of a diverse mixture of materials quite difficult to calculate, but in practice it is better to start by following general guidelines and gradually develop your own sense of the right amounts. One of the problems in trying to calculate the carbon–nitrogen ratio is that there is such variation in the analysis of the same product. For instance, poultry manure can vary between 3:1 and 10:1, depending on the freshness of the manure and the diet of the chickens. Some examples of typical carbon–nitrogen ratios: poultry manure, 3:1 to 10:1; fresh lawn clippings, 15:1; cow manure, 20:1; coffee grounds, 14:1 to 25:1; fresh weeds, 20:1; leaves, 30:1 to 85:1; straw/hay, 75:1 to 100:1; sawdust, 500:1.

Two other materials are of critical importance: air and water. Air is provided mainly via the provision of a straw core through the heap. All materials used should be moist: sprinklers can be set up to keep materials in holding areas moist before making the heap, and additional water can be added during construction if needed. Further details follow.

Manures

Any farm manures can be used. If you have a choice, be guided by the quality of the digestive process in the animal rather than the NPK content. Cows have the most thorough digestive process of all farm animals. Their manure is very refined already, thoroughly digested and high in microbes, making the perfect manure for composting. Poultry have a much less refined digestive process and their manure, though high in NPK, is less ideal for compost-making and is prone to overheating. More high-carbon materials (even a proportion of sawdust) are required, and a lower proportion of manure.

From my experience, I would rate the various farm manures in order of preference as follows:

1. Cow
2. Sheep/goat/alpaca
3. Horse
4. Poultry/rabbit/pigeon
5. Pig

Large quantities of manure can be obtained from cattle feed lots and poultry farms. Certified growers should consult their Biodynamic certifying body before bringing such manures on to the farm, and an initial test for chemical residues should be done if the source is not already on the radar of the certifying body. Periodic testing should be repeated every few years to ensure no contamination comes on to the farm. Experience has shown that most manures from conventional farms are clean; pesticides in animal feed tend to be stored in the fat or the liver.

If fresh animal manures are difficult to obtain in quantity, the high-nitrogen animal component can be provided through the use of commercially available products such as blood and bone or various composted manures. Fresh animal manure is of course best.

Dairy farmers have large amounts of manure available on the holding yard twice a day. This can be pushed off to a holding pit with a rubber-covered hand pusher, a rubber-tipped pusher mounted on a quad bike, or other mechanical pushers. After each yard cleaning, cover the manure with a good coating of hay or straw. When the pit is full, the manure is moved to the compost yard to await compost-making.

Beef farmers who want to make compost (whether to spread on fields or to use on a market garden run in conjunction with the beef operation) can also collect manure. Cattle can be fed hay in one area over winter (ideally on a concrete pad, but any dryish area will work). Periodically, the manure and hay on the ground can be pushed up and collected. The 80-hectare Australian Demeter Biodynamic method market garden, Agrilatina, in Italy, has an associated beef cattle property where a large amount of manure is collected for composting and use on the market garden. Ninety head of cattle are held at night on a partly roofed concrete feed pad on deep straw bedding. Extra straw is added regularly to balance the manure and urine build-up and to soak up excess rainfall, and the manure and straw are collected every 2 to 3 months and made into Biodynamic compost heaps using a side-throw mixer wagon: steel sheets are attached to the sides to direct the material into long windrows.

On sheep properties, large amounts of manure build up over time under the shearing shed and can be periodically cleaned out and composted.

Figure 13.1. Overnight cattle yard, Agrilatina, Italy.

The proportion of plant material to manure is largely determined by experience and direct sensing as you work. As a general rule, if your heap is made largely of cow manure and spoiled hay, you should aim for 30–50% manure and 70–50% hay by volume. If the manure is fresh, tend towards 30% manure; if older, tend towards 50%. If a lot of poultry manure is used, even 30% manure could cause overheating. If a lot of sawdust is used, then the manure content should be 60% or more. If your heap heats up quickly to 60°C or more, you have used too high a proportion of nitrogen-rich materials; these could include fresh green material, which is just as heating as manures. If your heap doesn't heat up sufficiently, you could make some liquid manure and pour this over the top of the heap. Remember to use more manure next time.

If you think that the materials you are using could cause overheating, some soil can be added to counter this, sprinkled throughout or used in thin (1cm) layers.

Materials other than manure and plant matter can be added to heaps – for instance, to address known mineral deficiencies in your soil. Materials that could be added to the heap include blood and bone, pelletised poultry or other manure (too much of these could cause overheating), rock dust (basalt), reactive phosphate rock, sulphate of potash, trace elements that are known to be deficient (very sparingly) and kelp powder from a clean source such as King Island (Australia). Don't become obsessed with additives or special ingredients. Compost of exceptional quality can be made from cow manure and hay only. The addition of lime or dolomite to compost heaps is often recommended. If used at all, they must be used very sparingly. I have never added either and am not in favour of this practice. There is a danger that nitrogen in manures can be leached out when manure comes in contact with lime. Well-formed colloidal humus has a pH close to 7 (neutral) without any addition of lime.

Human wastes

Rudolf Steiner cautioned against the use of human wastes in agriculture for various reasons, and their use is heavily restricted under organic standards. Material from composting toilets is best used in woodlots or similar, not for any crops intended for human or animal consumption.

Site Preparation

If possible, spray 500 at least once before building the first heap on the site. If it is a new site, all grass should be removed (this material can be used in the heap) and the site lightly cultivated to aerate the soil. If working by hand, chip the grass off with a hoe and work a fork back and forth in the soil to slightly loosen it. If preparing a larger site, for machine-making of the heap, mow or graze the site, then chip-hoe with a rotary hoe to cut the crowns of the grasses before cultivating with a tined implement if necessary, or just cultivate the grasses in. Be careful though. If the soil is already well aerated, cultivation could be unnecessary and make the soil too soft for heavy machinery, causing problems. If difficult grasses such as kikuyu or couch are present, make sure they are all killed by cultivating repeatedly during a dry, hot season, or removed before making the heap.

The area prepared should be larger than the expected size of the compost heap. For example, if the heap is to be 5m by 1.8m, clear a site 6m by 3m. Don't forget to run the heap north–south for even sun warmth. The minimum size for warmth retention and proper fermentation is about 1.5m³. This entails a heap whose base is 1.5m long by 1.2m wide and that tapers to a top of 1m by 0.5m and whose height is 1.5m. Farm-scale heaps can be up to 3m wide, 1.5m high and as long as desired.

Equipment

Handmade heaps

Wheelbarrow, hand fork, rake or drag fork, shovel or spade. Crowbar and/or sharpened stakes to make holes for the preparations.

Large-scale heaps

Front-end loader, manure spreader (Krone or similar) or mixer/feedout wagon. Also, a

Figure 13.2. Grass removed and soil loosened and a straw A-frame placed along the centre line of the heap.

second tractor, to run the spreader while the front-end loader adds material to it. Crowbar or sharpened stakes as above. Alternatively (but less satisfactory), mixing can be done by laying out materials in a windrow and running a rotary hoe over the materials, set so as to not penetrate the soil surface, then collecting the mixed materials and adding them to the heap with a front-end loader.

A Krone manure spreader (or modified mixer/feedout wagon) is well suited to compost-making. Chains and bars on the floor of the manure spreader move material back to a series of sharp beater blades that chop and mix the materials before throwing them up and outwards. This is ideal for spreading finished compost on fields. For compost-making, a

steel cowling is fitted over the beater blades. Material that is thrown up by the blades hits the cowling and falls down, forming a heap that naturally tapers towards the top. When the heap is around 1.5m high, the operator moves the spreader forwards so that a long compost windrow is progressively created. This can be as long as desired.

If you are using dairy manure, covering each milking's collection with a generous layer of hay or straw creates a good mixture of materials that will be thoroughly combined by the manure spreader when the compost is made. Some extra hay or straw can be added as the spreader works, if more is needed at compost-making time. For instance five or six extra small square bales can be added on top of a load using the front-end loader.

Handmade Heaps

A typical small handmade heap will be 1–1.5m wide at the base, 1.5m high and as long as you have material for. Remembering that the heap should run north–south, make a central core of straw books[1] in pairs to form a line of A-frames running the length of the heap. The purpose of this core is to allow air to penetrate right through the centre of the heap. Without it, the materials in the centre will not convert into colloidal humus. Some people run a bundle of sticks and thin branches, or a length of 100-mm slotted drainpipe, under the A-frame to enhance aeration. The straw A-frames tend to gradually settle down under the weight of compost and breathe less freely; the bundle of sticks or slotted drain pipe keeps the centre a bit more open. This is not essential but can be

Figure 13.3. Starting to mix materials.

Figure 13.4. The mixed materials have nearly reached the top of the A-frame. At this stage, tidy up the sides to begin to form a steady, fairly steep slope.

Figure 13.5. Progressing – now the top needs to be levelled and the sides tidied before continuing.

Figure 13.6. Finished heap ready for insertion of Biodynamic compost preps.

helpful. Don't cover the ends of the A-frames, sticks or pipe with compost material. Leave them open to allow air flow.

The best results in compost-making are obtained if the manure and plant materials are thoroughly mixed before being added to the heap. Mix close to the ends or sides of the heap. If you are using a 50/50 mix of manure and other materials, put equal quantities of each on the mixing site and stir them up with a garden fork until they are fairly well mixed. Put the mixed materials around the straw A-frame on the heap

site. Move more materials to the mixing area and repeat the process.

Take care not to dislodge the straw A-frames in the early stages. Keep working until you have covered the area of the base of the heap to about 200mm deep. Then, using a rake or fork, tidy up the base to form a neat rectangle. Keep adding mixed materials until you have reached an ideal height of 1.5m or a minimum of 1.2m. As you work, keep tidying the walls so they form a steep gradient, by pulling material towards the edges with a rake or drag fork. Keep it neat.

If you don't keep the sides fairly steep, the heap will narrow off too quickly and will not reach the desired height. Adding material afterwards doesn't help; it falls off. Too vertical is also not good, since the final insulating layer of hay or straw will not stay in place. If you run out of material before you reach a height of at least 1.2m, you have made the base too large for the available material; break material off one end and add it to the top. You will develop a better sense for quantities and size with experience. In shaping the heap, bear in mind that in summer it is best to have a wider, slightly concave top to assist watering, and in winter a narrower, convex top to better shed excess rain.

Although thorough mixing of materials gives the best results, it does entail extra work, which is difficult for those who are elderly, infirm or suffer from a bad back. An alternative method is available that produces just as good a result: thin layering. This method involves layers of absolute minimal practical depth – no more than about 4cm. Manures and plant materials are placed in alternate thin layers and raked to spread them evenly. The cumulative pressure of successive layers tends to mesh these layers well, and good compost results. Layers any thicker than about 4cm will not work nearly as well. This method has proven successful in many trials over the last 40 years.

Larger, Machine-Made Heaps

Heaps made with machinery are wider than a handmade heap, up to 2.5 or even 3m. They require a larger central core to ensure that adequate air reaches the centre of the heap. This can be achieved by, for instance, running a line

Figure 13.7. Making compost on a Biodynamic dairy farm using a Krone manure spreader – Mark and Lynne Peterson, Nathalia, Victoria. Photo: Mark Peterson.

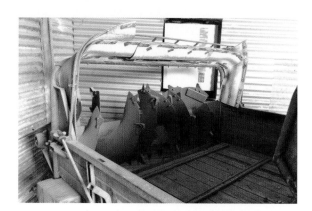

Figure 13.8. Manure spreader showing blades. Cowling off.

of whole straw bales (straw bales breathe better than hay bales) end to end under the centre line of the heap, or even running two lines of straw bales leaning against each other to make a continuous A-frame.

Moisture

The moisture level of compost materials is critical to success. Too dry and decomposition will be slow and inefficient. Too wet and the heap becomes anaerobic. If materials have been left outside exposed to rain for a time before composting, they should be moist enough. The material should feel moist and if you can squeeze a few drops of water from them, that is good. Otherwise, the heap should be hosed with water during construction to make sure it is uniformly moist.

The Biodynamic Compost Preparations

The six Biodynamic compost preparations, numbered 502–507, are added to the heap when it is completed, to regulate and enhance the composting process. The first five are solids, whereas valerian (507) is a liquid that comes in a small glass test tube. The actual amount of each solid preparation per set is only 2g. If this small amount, about the size of a pea, were to be posted as is, the preps would be almost completely ineffective. The solid preparations must retain their moist, colloidal nature (the oak bark preparation is not properly colloidal but must too retain its moisture). Unless you are making large amounts of compost, in which case you will have a special compost preparation storage box, and weigh out the sets of preps yourself, they are posted with each solid preparation encased in its own golf-ball-sized ball of colloidal humus (good compost or 500) to protect them. If you are weighing the preps yourself, similarly encase each one (the solid preps) in a ball of colloidal humus before

Figure 13.9. Preparation storage at Agrilatina, Italy: compost prep box to left, 500 box to right. Photo: Agrilatina Cooperative.

inserting it in the heap. If you are storing large amounts of compost preparations, a special box similar to the 500 storage box is required. Six separate wooden compartments are made, each separated by 80mm of peat moss, and with 80mm of peat moss around the outside of the compartments and in the lid. The five solid preps are put in small glass jars with a lid left loose, one to each compartment, valerian in a sealed jar in the sixth (or stored in a sealed bottle on its side in a cool dark place). The box is stored in a similar place as with the 500 storage box, away from petrol fumes and electricity.

The compost preparations may be added as soon as the heap is built. However, many compost-makers prefer to wait until the initial heating phase is over and the temperature has declined to around 40°C or less before adding the preparations. This is also my preference.

Handmade heaps

For handmade heaps, the preparations are inserted in the heap so that they are in the centre

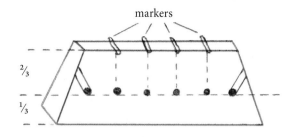

markers

2/3

1/3

Figure 13.10. Markers guide even spacing of preparation holes.

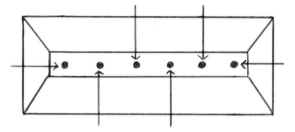

Figure 13.11. Arrows show on which side the holes are inserted.

of the total mass of material. As there is more mass at the bottom than at the top, the preps should be placed at about a third of the height of the heap, ensuring, however, that they are well above the top of the straw core. There is no set position for any of the preps; the important thing is that they each have their own individual space to work in. For small heaps the liquid valerian preparation is normally placed in one of the end holes.

One preparation goes in either end of the heap and two are put in from each side. Place four markers (e.g. twigs) on top of the heap, equally spaced, ensuring that the two towards the ends are far enough away from the ends that the preps placed directly under them will not be in the same position as the end preps. Holes are pushed into the heap at the appropriate positions at a 45° angle (or steeper)

downwards. See Figure 13.11 for the direction to make the holes from.

Various implements can be used to make the holes. Initially, use a crowbar, rake handle or garden stake to open the hole, then use a thicker pointed stake (60–75mm square is ideal) to enlarge the hole. The hole must be big enough for your arm to go in. Having made the six holes, place a nest of old compost or good garden soil 50–75mm thick at the base of each hole. When each nest is in place, put solid preparations in five of them, making sure each goes right to the bottom of its hole; reserve one of the end holes for the liquid valerian. If your preps are wrapped in unbleached waxed paper, don't forget to take it off.

Fill each of the five holes with old compost or good garden soil (non-heating materials), pressing it in firmly until level with the wall of

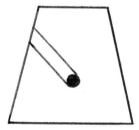

Figure 13.12. Side hole.

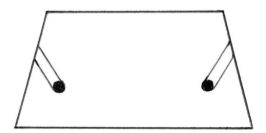

Figure 13.13. End hole.

Figure 13.14. Making an initial prep hole with a crowbar.

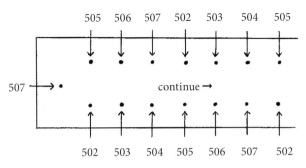

Figure 13.15. Example of preparation distribution in a large, machine-made compost heap.

the heap so that no depression will later form that could allow water to collect. Surrounding each preparation with old compost or soil (non-heating materials) helps keep it cooler than the rest of the heap.

The valerian liquid is poured into 1 imperial gallon (4.5 litres) of pure warm water and stirred as for 500 for 20 minutes. An earthenware crock or stainless steel pot or bucket is ideal for this quantity of water. After stirring, pour half the liquid into the last hole and sprinkle the other half evenly over the heap. One of the important functions of valerian is to attract earthworms to the heap. Fill this last hole with soil or old compost after the liquid has soaked in.

Machine-made heaps

The compost preparations are inserted as for handmade heaps, with one important difference. Owing to the extra width of the machine-made

heap, two lines of preparations are inserted rather than one down the centre. The preps are inserted from each side so that the two lines are about 1m apart (or so that each prep is roughly equidistant from the side of the heap and the centre line), so that each prep has a generous space in which to work independently. The preps in each line are also spaced about 1m apart, adjusted for even separation according to the overall length of the heap. Each set of preparations (2g of each solid, 5ml of valerian liquid) is enough for 10–12m³ of compost. Place each subsequent set of preps in the same order so that each type is spread evenly through the heap.

Covering the Heap

A thick layer of insulating material is spread or laid over the heap, to retain moisture and protect from wind and sun. This can consist of materials such as straw, hay or reeds. Straw books (sections broken off a bale, 50–75mm thick) can be laid like tiles, or the material can be loose. Do not use soil or any other material that does not breathe well. If you are using loose material, the covering should be at least 100mm thick,

Figure 13.16. Heap covered with loose material.

Figure 13.17. Heap covered with straw book 'tiles'.

preferably 150mm, and will need replenishing as it is gradually digested or weathered. Hosing down well after application helps the material hold together and not blow off. If the heap is well insulated, the compost will become colloidal right through to the outside. If it is not well covered, the outer parts will dry out and not convert to humus.

If you are inserting the preparations as soon as the heap is built, the covering remains until the compost is ready to use. If, however, you wait until the temperature falls below 40°C, you will have to pull the covering off the top half of the heap so you can prepare the holes for the preparations, and then replace the covering.

Monitoring the Heap

The temperature of the heap should be monitored right from the start. If you have used too low a carbon–nitrogen ratio (i.e. too much nitrogen) the heap may become too hot within a day, releasing nitrogen in the form of ammonia gas. If the temperature goes over 55°C, water the heap from the top to bring it down. Obviously, if you have to water often to reduce the temperature, the heap may become too wet and valuable nutrients will leach out. Next time, use a higher carbon–nitrogen ratio.

If the temperature refuses to rise above 40°C, make a liquid manure (or liquefy some blood and bone or similar) and pour this over the heap. Next time use a lower carbon–nitrogen ratio.

A glass scientific thermometer or a special long, metal-spiked digital thermometer can be inserted well into the heap to check temperature. A scientific thermometer can be protected from breaking by putting it in a 15mm copper pipe that is long enough to contain the whole length of the thermometer. One end of the pipe is flattened with a hammer so that the thermometer can't slip out. The pipe is pushed into the compost heap to its full depth. Leave for

Figure 13.18. Checking the compost, Agrilatina, Italy. Photo: Agrilatina Cooperative.

Figure 13.19. Sprinklers on top of compost heaps, Chislett's Biodynamic citrus grove, Victoria.

a minute or so, then remove and read it quickly, for the temperature reading will drop rapidly once the thermometer has been removed. Never use mercury-filled thermometers in case they break. In the absence of a thermometer, judge temperature with your hand. If you can hold your hand in the heap for 5 seconds or so, but no longer, the heap is close to the maximum desirable temperature. If it is too hot to hold your hand in at all, it is definitely too hot.

Moisture content should also be monitored, and more water should be hosed on from the top and sides if necessary. For farm-scale heaps, a line of sprinklers can be set up along the top of the heap. Some natural rain on the heap is good, but prolonged rainy periods can make the heap too wet. A plastic tarpaulin can be put over the heap temporarily, making sure it doesn't cover the whole heap and leaves the lower parts of the sides open. Remove the tarpaulin after the excessively rainy period ends.

When Is the Compost Ready?

The compost is ready when it is seven-eighths digested and has the texture of colloidal humus – that is, you can mould it in your hands like putty. You may be able to see the outline of a piece of straw or similar material, but it disappears when kneaded. Otherwise, ingredients can't be identified. The compost should be sweet smelling and shouldn't stain the hands.

It is best to use the compost while the worms are still active, since then it still retains maximum nutrient value, but in a home garden, where the compost is used bit by bit as required, this won't be possible. Always start at one end of the heap and continue removing from that end, covering well in between times.

Provided the heap is kept well covered and watered in dry periods, good compost can retain its colloidal nature for a year or more.

It is not possible to put a set time on the composting process. Depending on the site,

Figure 13.20. Moist, colloidal humus compost, Buninyong, Victoria.

the quality of the materials, and other factors, it could take from 3 to 6 months to make good colloidal humus; longer if too many aspects are not ideal. As each successive heap is made on the same site, the time to completion reduces until the site reaches its potential.

How to Use the Compost

Compost is best worked into the soil if it is to retain its qualities and feed plants most effectively. If it is left on the surface it can dry out and lose its colloidal nature. Place a layer on the soil and turn it under with a garden fork before sowing or planting. Or dig a shallow trench, add

Figure 13.21. Spreading compost, Petersons' dairy farm, Victoria. Photo: Mark Peterson.

compost, turn it under and rake the soil back over the top. On a market garden, work it in with tines before bed-forming.

When applying compost in situations where cultivating is impractical, such as on pasture, in orchards or on berries, a different approach is needed. Apply to pasture in spring after grazing and when grass is actively growing. Irrigation after application can help. The roots will take in the compost before it dries out. The same applies to fruit trees and berries that are grown with grass cover. Where mulch is used, compost can be applied to the soil in spring and mulch applied to protect it.

How much to use depends on your soil fertility and the crop to be grown. For a heavy feeding crop (e.g. cauliflowers) in a poor soil, I would use a 50mm layer of compost, turned in; less in a more fertile soil. For the next crop, which would be a light feeder, a legume or a root crop, there may still be sufficient humus, though a small addition may be necessary in a poorer soil. It is a matter of sensing the needs of the soil and the crop.

This sensing ability is a most important aspect of Biodynamic development. It comes with experience and the use of active perception (see p.49). Are the plants looking poor and hungry, are they too big and gross (overblown) or are they actively growing with finely structured features and a 'glow-green' colour? Am I using too little compost, too much or just the right amount?

I find that if I use the right amount of compost, heavy feeders (e.g. celery, cauliflowers) will produce well without the need for any extra feeding (e.g. with liquid manure). The compost will continue to feed the plants right through the growing season.

Biodynamic Sheet Composting

Biodynamic sheet composting is an in situ form of composting, in conjunction with the use of prepared 500 when appropriate conditions apply. Sheet composting includes:

- Dynamic rotational grazing of pasture, in conjunction with the harrowing of cow manure after grazing.
- Topping of pasture, mainly in spring.
- Slashing or mulching (with a PTO [power take-off] mulcher) of pasture growing in orchards, vineyards and berry plantings.
- The growing and working in of green manure crops. Raw manures and other inputs can be applied before sowing, provided two green manures are grown in succession (the second without any manure added) before the growing of a food crop.

These forms of sheet composting will be detailed in Chapters 17 and 18.

FOOTNOTE FOR CHAPTER 13

1.   Compressed slabs of straw formed in the baling process.

Chapter 14

Biodynamic Liquid Manure

The primary focus in Biodynamic plant nutrition is the building up of a bank of soil colloidal humus, enabling plants to grow and feed naturally under the guidance of sun warmth and light. Provided adequate humus is present, healthy crops can be grown without any additional feeding.

For nearly 50 years I have grown my own vegetables Biodynamically – and for a time I ran a Biodynamic market garden, with partners. I have never found the need for any form of supplementary feeding, provided that I have built adequate colloidal humus before planting or sowing, through either green manuring, compost application or a combination of these. Heavy feeders such as celery or cauliflower were well fed solely by the pre-planting humus supply. However, there *are* situations where supplementary feeding may be useful, and Biodynamic liquid manure can be used on a growing crop without force-feeding. These situations generally involve insufficient colloidal humus being available, for whatever reason, for a particular crop.

- Fill a container such as an old bathtub or clean 44-gallon drum (not plastic, or one that has stored chemicals) a quarter full with manure. If you can use a variety of different manures, so much the better.
- Fill the container with pure water.
- Break up the manure and mix as much as possible.
- Make a wooden cross big enough to sit over the top of the container and hang a set of the five solid Biodynamic compost preparations from the cross as shown in Figure 14.1 (below). The preparations (each encased in a golf-ball-sized ball of colloidal compost or 500) are tied in open-weave natural cloth bags (muslin/cheesecloth is ideal). A stone in each bag will stop the bags floating to the surface.
- Make the strings on the bags long enough that the balls hang a quarter to a third of the way down in the liquid.

Figure 14.1.

- Use the preparation balls as supplied by your preparation supplier, or make your own balls using 2g of each solid preparation encased in a golf-ball-sized ball of colloidal compost or 500. It doesn't matter which preparation goes in each position on the cross. Sit the cross over the container.
- Stir the liquid valerian preparation (5ml) in 1 imperial gallon (4.5 litres) of warm water (as for 500 – see p. 71, 73–74) for 20 minutes, then pour it into the container.
- After temporarily removing the wooden cross, stir the container thoroughly twice a week. The best implement for this job is a cream stirrer or similar. You can make your own using a 150–200mm-diameter circle of galvanised metal or heavy-grade plastic, with 19–25mm holes drilled into it, and a long handle attached (an old spade or fork handle is ideal). See Figure 14.2. Use a vigorous up-and-down plunging action to break up the manure and mix it well.
- Use when the liquid develops a black, oily, glazed appearance. This may take 2–4 months depending on the time of year. When the liquid reaches this stage it has a somewhat colloidal nature and will not overfeed plants. Dilute it 1:4 or 1:5 with water before use, but be guided by experience and take care that plants don't develop an overblown appearance from overuse.

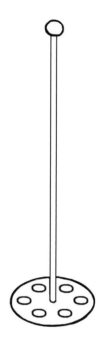

Figure 14.2. 150–200-mm-diameter metal plate with 19–25-mm holes and long handle.

Chapter 15

Supplementation of Biodynamic Soils

When converting a farm to Biodynamics, it is important to handle the transition carefully. Prepared 500 will powerfully and progressively transform soil in terms of biological activity, root growth and humus formation, leading to deeper and better-structured soil. In better-quality soils there may be no need for supplementation with minerals. But in many cases there will be a need for supplementation, at least in the early stages, and quite often this may be ongoing. The two main aspects of a balanced soil are good biological activity and sufficient elements for plant growth. These go hand in hand. Alex Podolinsky wrote,

> The feeding of elements and the stimulation of biological activity must be attuned to each other. That is the basis of true organic farming … we must watch that in feeding elements we do not go beyond the time factor of the biological development of the soil, or that we do not feed insufficient elements either, so that although the biological activity is stimulated, the soil does not grow anything.[1]

When it comes to plant feeding, there is a tendency to focus only on the soil. However, the plant lives and interacts not just in soil but also in a much wider context. All the elements are present in air, many in minute amounts, and can be absorbed by leaves. Moreover, the leaf creates new substances via sun, water and carbon dioxide through photosynthesis. Legumes fix nitrogen from the air through the activity of soil bacteria called rhizobia. The soil balance and the elements available to plants are a complex interplay of many factors, including soil biology, the plants themselves and the wider environment.

For the first 30 years or so of Demeter Biodynamics in Australia, many farmers aspired to the ideal of the closed farm unit – in other words, using no fertiliser inputs at all. Podolinsky maintained high production on his dairy farm at Powelltown for 40 years without any fertiliser inputs. Years of below-average rainfall eventually necessitated an application of reactive rock phosphate. It was gradually realised that production on many Biodynamic farms that used no inputs was falling over time. The nature of modern low-labour large-acreage farming means that large amounts of produce need to be exported year after year in order to make a living, which makes it much harder to maintain a closed unit than on European organic farms of the past, which recycled everything and exported a much lower percentage of their production, much of their produce being required to feed their large workforces and families. Additionally, Australian soils are very old geologically and weathered, many having deficiencies in elements. The generally poorer soils and erratic Australian climate are much less conducive to a closed system than conditions in large parts of Europe and North America.

In general, the input requirements on Australian Demeter Biodynamic method farms are much lower than those of conventional farms, and also noticeably lower than those of other organic farms. The enormously enhanced biological activity engendered by the ongoing use of prepared 500 and the ability of living systems to biologically transmute elements – especially effective in biologically active soils, rich in humus and microorganisms – gradually brings soils into balance.

In some situations, once initial deficiencies have been addressed and biological activity has been developed through the use of prepared 500, no more elements will need to be applied for many years, even decades in some cases.

Determining Soil Needs

Observation

Any Biodynamic grower, whether a gardener, smallholder or farmer, needs to develop keen observational skills. Learning to assess soil structure will help inform you about your soil's biological development. Observing plant growth will provide much useful information about the nutritional requirements of plants, as will pasture

composition, cropping levels, animal health, carrying capacity and many other factors. This takes experience and the ability is developed over many years. An experienced visiting Biodynamic farmer can often pick up many clues and provide very useful advice to a novice.

Soil tests

Soil tests for major and minor elements, cation exchange capacity, organic matter levels and other factors can be very useful, but need to be viewed with caution, particularly with regard to any ensuing recommendations.

Soil is a living, evolving entity. Soil elements are in constant flux, changing from season to season, month to month, even day to day. A soil test is a snapshot in time, though it can provide useful indications. In the early 2000s, the Australian Soil and Plant Testing Council conducted a survey on Australia's soil-testing laboratories, sending identical standardised soil samples to 18 laboratories across the country. The results called into question the reliability of soil tests: the results for nitrogen, phosphorus and potassium varied widely between laboratories; phosphorus levels in one sample varied between 15.3 and 56mg/kg, and potassium levels varied between 70 and 600ppm. No adequate explanation of these discrepancies was suggested, but it is possible that biological transmutation of elements under different warmth, humidity and other factors, including ongoing biological activity in the soil samples during transit to the laboratories, could partly explain the discrepancies.

Testing laboratories often provide recommendations for fertiliser application on the basis of test results. These are based on a theory of plant requirements that is still in its infancy and doesn't take into account some important factors, including the pivotal role of soil humus. Soil pH is a case in point. Huge amounts of lime are often recommended to bring the soil pH up to an 'acceptable' level for a particular crop. In a high-rainfall area, low pH is, in part, nature's way of preventing soil elements becoming too readily available and prone to leaching. If a large amount of lime is applied at once, large amounts of elements can suddenly become available and be lost from the soil. Colloidal humus is chemically almost neutral, with a pH close to 7. In a Biodynamic soil with a low pH, but good humus levels, 'lime-loving' plants can still thrive. Moreover, it has been the experience of many Biodynamic farmers that low soil pH increases steadily over time, even without the application of lime; by contrast, conventional growers generally experience increasing acidity over time. In general, Biodynamic farmers are recommended to apply lime at a maximum rate of 200kg per acre (500kg per hectare) at a time. If more is needed, it is better to apply small amounts periodically rather than a large amount all at once.

In many situations, soils gradually come into balance with the regular use of prepared 500, through the development of biological activity, higher humus levels and open, friable structure, without the need for inputs. It takes time to learn when and how to remedy deficiencies. A combination of soil test results, careful observation, and advice from experienced Biodynamic farmers, particularly those in your geographical area, should inform decision-making.

Fertility drift

Farmers should take into account the fertility drift that often occurs on farms: fertility tends to become lower on hills, higher in valleys, and higher towards stockyards or the dairy on dairy farms, and lower away from the concentration points. For instance, hay can be fed on higher ground (often advisable in wet times anyway) to reverse the trend. On farms where compost is made, it can be applied to the poorer parts of the farm first, and at higher rates than on the more fertile areas.

Fertiliser Inputs in Biodynamics

In general, certified Biodynamic growers can use any fertiliser inputs that are listed as allowable inputs in their country's organic/Biodynamic standards, with a few caveats. Professional growers new to Biodynamics are advised to consult their national or state Biodynamic association and, ideally, a local mentor as to what products are best to use in each situation, and at what rates.

Compost and green manure

Biodynamic compost is an ideal way of adding elements to soils, in the form of colloidal humus that is easily taken up by plant fine hair feeder roots under the direction of the sun. Any imported compost materials should be checked as described on pp. 92.

Green manure crops are also a very good way of providing necessary elements in the form of colloidal humus, once the worked-in green manure has had time to fully break down. Green manure crops may well require fertilising with any of the inputs listed below, which can also be applied to soils by themselves or in combination.

Rock dust

A very good source of (insoluble) elements, including trace elements, which will slowly become available through soil biological activity. Different forms are available, including straight basalt rock dust (crusher dust) and various rock dust blends. Rock dusts can be used in relatively small amounts or up to 1.5 tonnes or so per acre (up to 3.75 tonnes per hectare), depending on the situation. Other rock dusts include those derived from granite, rhyolite, diorite and andesite. Each has different levels of minerals.

Reactive phosphate rock (RPR)

An insoluble form of phosphorus, which is released slowly by reaction with acid soils together with the soil biology. Contains around 13% phosphorus. It is most suitable in acidic soils, and results can be variable. An observable response may take several months, and the RPR will continue to release phosphorus for several years. It can be used at rates varying from 50 to 500kg per hectare. Some rock phosphates contain high levels of cadmium and should be avoided.

Soft phosphate rock

This is a more soluble form of phosphate, suitable for soils that are neutral or alkaline, where RPR is not successful. Can be used at similar rates as RPR, but small amounts at different times can be better than a large amount at once.

Potassium

Potassium deficiency is less common than phosphorus deficiency. Various minerals can be used to address potassium deficiencies. Non-water-soluble forms such as white mica (muscovite), black mica (biotite) and greensand (glauconite) are preferred, but, if these prove hard to source, sulphate of potash (water soluble) may be used to address measured or observed potassium deficiencies and for potassium-hungry crops. The normal recommendation is to use sulphate of potash sparingly, around 5kg per hectare. However, for a crop with a high potassium need (and where potassium is not already well supplied), such as potatoes or bananas, or in high-rainfall tropical areas, it can be used at rates as high as 125kg per hectare or more on some crops in some situations. Consult your national Biodynamic association for advice.

Lime and dolomite

As explained earlier, colloidal humus has a pH close to 7, and acid soils well supplied with colloidal humus can successfully grow plants that are normally regarded as needing a higher pH. However, lime or dolomite – use dolomite where there is a magnesium deficiency – can be applied periodically if needed, generally at no more than 500kg per hectare. More can be used in some situations, but remember that damage can be done and soils can be unbalanced by freeing up too many elements at once through large applications.

Gypsum

Gypsum (calcium sulphate) is another form of calcium and also a source of sulphur. It provides an additional benefit in clay soils by flocculating fine clay particles into aggregates, temporarily improving soil structure and drainage and replacing sodium on the surface of the clay particles with calcium, thereby allowing salt to leach out of the soil. Soil structure improvements only last a few months but this window can be used to allow actively growing roots to build soil structure in conjunction with 500 application, working in a green manure or similar activities to produce longer-lasting improvements in structure. Gypsum can be used at 125 to 1,500kg per hectare, depending on the situation and crop (more in certain circumstances), and tending towards the higher end of this range on heavy clay soils.

Guano

A natural fertiliser derived from varying sources, depending where it is mined. The guano used in Australia derives from the remains of coral, seabird droppings, fish and seaweed deposited over thousands of years. It is a valuable source of citrate-soluble (not water-soluble) phosphorus (13%), calcium (33%) and silica (24%) together with a very low nitrogen content (0.1%) and a good range of trace elements. It can be used at around 125kg per hectare, more if necessary.

In some countries, seabird or bat droppings make up a higher proportion of guano, resulting in much higher nitrogen levels, as much as 10%. Care should be taken not to excessively push plant growth with such a high-nitrogen content substance, though it would be fine for application to a green manure crop.

Fused calcium magnesium phosphate (FCMP)

Apatite is heated and fused with calcium and magnesium silicates to form FCMP, a non-water-soluble supplement. Low-cadmium sources are available in Australia.

Sulphur

Sulphur can be used if a known deficiency exists. If calcium is also lacking, use gypsum. Prilled sulphur is an important lick component for sheep (for wool production).

Trace elements

These can be applied if a deficiency is discovered via a soil test or if the area is known for a deficiency in any trace element. Use natural chelates or non-chelated forms from rock dusts, not synthetically chelated trace elements. Consult the BDRI in Australia or the equivalent certification body in your country if you are a certified grower. The Australian Demeter Biodynamic standard does not allow foliar application of trace elements (or any other fertiliser) by certified growers, since this circumvents the natural humus feeding of plants under the guidance of the sun.

Examples include: zinc oxide, a non-water soluble alternative to zinc sulphate; natural magnesium carbonate (magnesite, 28% magnesium); natural magnesium oxide (periclase, 60% magnesium); and calcium magnesium boracite (hydroboracite, 10% boron, 16% calcium and 6% magnesium).

Seaweed products

These are available in pure form, as powders or liquids, and also blended with various manures in pelletised form. They provide a wide range of elements, including trace elements. They provide water-soluble elements and are therefore best used in compost heaps, on green manures before a crop, or on pasture before a cropping phase. The area where the seaweed was gathered needs to be considered, since some areas may be contaminated by industry.

Seawater concentrates

Seawater concentrates are a promising recent development that is well suited to organic and Biodynamic agriculture. By removing 95% of the sodium chloride content and then removing most (in some products, all) of the water, the remaining roughly 90 minerals are concentrated, together with the active organic components.[2]

Around a dozen groups worldwide are developing seawater products as fertilisers and as animal feed supplements. In the US, seawater concentrate is marketed as Seacrop® or SeaAgri.[3] In Australia, Moodie & Associates[4] have developed a process whereby pristine, cold seawater containing high levels of plankton is collected from the southern Australian coast and refined ashore in a three-stage process. The end result contains all the non-sodium and non-chlorine minerals concentrated to 2.5% of the original mass of seawater and includes 0.5% biota.

Manures and animal-based products

These should be treated with caution, since they contain water-soluble elements that will feed plants indiscriminately. Even if called a 'composted' product, they are generally not in colloidal form and can therefore become water soluble in the soil. They can be used in Biodynamic compost heaps or directly on green manure crops ('sheet composting').

If raw manure is applied to a green manure crop, a second green manure crop should be grown after working in the first, without further addition of raw manure, applying prepared 500 at appropriate times, before a food crop is grown. Cow manure can be applied to green manure crops at around 3 tonnes per acre (7.5 tonnes per hectare). Manure and animal-based alternatives for green manures include blood and bone and pelletised manure products. Apply 200–400kg per acre (500–1000kg per hectare).

Small amounts of pelletised manure products or blood and bone may be applied to fruit trees or berries, provided this is done after cropping is completed, ideally in conjunction with a green manure, allowing adequate time for the soil life to incorporate the soluble elements in the soil humus before a new crop begins.

Certified growers are advised to consult their national certifying body (the BDRI in Australia) before importing any raw manures, manure-based pellets or blood and bone. Periodic testing of these products for chemical residues may be necessary. A comprehensive list of permitted materials for soil fertilising and conditioning from the (Australian) National Standard for Organic and Bio-Dynamic Produce appears in Figure 15.1. (overleaf).

FOOTNOTES FOR CHAPTER 15

1. Podolinsky, *Biodynamic Agriculture Introductory Lectures Volume 1*, pp. 25–26.
2. A very good overview of seawater concentrates is Arthur Ziegler, *Seawater Concentrate for Abundant Agriculture*, Ambrosia Technology, 2012.
3. Marketed in Australia by Green Acres Biodynamics.
4. See Moodie & Associates, *Seawater Concentrates for Australian Agriculture*, 2021. Email: tony@moodieEngineering.com.au

Appendix B - Permitted materials for soil fertilising and conditioning

Table A1 Soil fertilising and conditioning

Substances	Specific conditions/restrictions
Animal manures	Application must be composted or followed by at least two green manure crops in cropping system.
Blood and bone, fish-meal, hoof and horn meal, or other waste products from livestock processing	Following application, uptake of such products by livestock does not form part of the animals' diet.
Compost	Should be produced in accordance with Australian Standard 4454-1999 or recognised equivalent system.
Minerals and trace elements from natural sources, including: • calcium (dolomite, gypsum, lime); • clay (bentonite, Kaolin, Attapulgite); • magnesium; • phosphate (rock phosphate, phosphatic guano); • potash (rock & sulphate potash); • elemental sulphur.	Must not be chemically treated to promote water solubility
Epsom salt (magnesium sulphate)	None
Microbiological, biological and botanical preparations	Products derived from genetic modification technology are prohibited
Mined carbon-based products	Peat to be used for plant propagation only
Naturally occurring biological organisms (e.g. worms) and their by-products	None
Plant by-products	From chemically untreated sources only
Perlite	For potting/seedling mixes only
Sawdust, bark and wood waste	From chemically untreated sources only
Seaweed or algae preparations	None
Straw	From chemically untreated sources only
Trace elements & natural chelates, e.g. ligno sulphonates & those using the natural chelating agents e.g. citric, maleic & other di-/tri-acids	Not synthetically chelated elements
Vermiculite	For use in potting/seedling mixes only
Wood ash	From chemically untreated sources only
Zeolites	None

National Standard for Organic and Bio-Dynamic Produce – Edition 3.8

Figure 15.1. Appendix B, Table A1.
National Standard for Organic and Bio-Dynamic Produce, Edition 3.8: November 2022, Department of Agriculture, Fisheries and Forestry, Canberra. Reproduced here under a Creative Commons Attribution 4.0 International Licence. No changes have been made. Copyright is owned by the Commonwealth of Australia. Disclaimer: The Australian Government acting through the Department of Agriculture, Fisheries and Forestry has exercised due care and skill in preparing and compiling the information and data in this publication. Notwithstanding, the Department of Agriculture, Fisheries and Forestry, its employees and advisers disclaim all liability, including liability for negligence and for any loss, damage, injury, expense or cost incurred by any person as a result of accessing, using or relying on any of the information or data in this publication, to the maximum extent permitted by law.

Chapter 16

Cultivation

Cultivation goes hand in hand with the Biodynamic preparations as a key developer of soil structure and fertility. Done well, it can improve or at least maintain the open friable structure developed with the use of Biodynamic preparations. Done badly, it can ruin soil structure and compact soils. Cultivation had a highly revered status throughout millennia of organic agriculture before the advent of 'conventional' agriculture.

At the dawn of settled agriculture, it has been suggested that the earliest Zarathustra (there appear to have been several over a long period of time), a personage associated with the development of agriculture, first suggested ploughing 'to bring light into soil'. In traditional organic agriculture, farmers were intimately connected with the soil and regarded ploughing as a deeply reverent act, one to be given the utmost care and attention. They instinctively understood how important soil structure was to soil and plant health. In medieval Wales, one farmer would walk backwards in front of the oxen, singing to them, while another guided the plough. The ploughman's eye would be intently focused on the soil (instinctive *active perception*) as the plough ran through it, sensing, adjusting carefully to suit the best development of the soil. The already well-structured soil was gently broken along natural fracture lines, rather than pulverised into dust as is so often the case today, when massive tractors race to get the job done with little care and often inappropriate implements. Ploughing was done in several stages, with resting times in between, the farmer watching carefully to determine the best time to work the soil again. The result was the maintenance of beautifully structured, fertile soils.

Alex Podolinsky described how, in his youth, in areas in Europe where traditional organic agriculture was still practised, children would be severely punished if they ran on to a hay paddock, because their feet would cause noticeable damage to the soil structure. If a car drove on to such a paddock, it would leave wheel ruts 100mm deep. The loss of soil depth on the magnificent Ukrainian soils (the best in the world) has been measured to be around 1m. This is the result of the destruction of structure and air pockets by conventional agriculture.

Compaction

One of the first steps when beginning Biodynamics is to dig in the soil to determine whether there is compaction. Preparation 500 develops and structures soil in conjunction with plant roots, working deeper and deeper. However, soil biology can't develop in a compacted soil, which lacks air, and roots cannot easily penetrate compacted layers (hardpans). If the soil is compacted, deep ripping can remedy the situation. This should be done when the soil is semi-dry. Ripping wet soil merely smears the cut surfaces and doesn't deal with compaction. If the soil is very dry, large blocks of soil may be brought up. In semi-dry soil, a ripper will relieve compaction while breaking the soil into more natural pieces. Once compaction is relieved, it is important to get plants growing so that their roots can begin building structure and humus, ideally in conjunction with prepared 500 if conditions allow.

Hand Digging

Let's look first at hand digging, since the principles for hand digging can be extended to larger-scale cultivation, and the same degree of care is required. Firstly, we need to be careful when we cultivate, soil moisture being the most important consideration. Never cultivate when the soil is too wet; we can soon see if this is the case once we start. I have, once or twice, when pressure was on to start a spring crop, begun ploughing when the soil moisture was borderline too high. I stopped after a few metres and got off the tractor to check, realised it was too early and went back to the shed. I could see the adverse effect on that section for months afterwards. As noted above, over-wet soil becomes smeared over rather than naturally broken. It is too wet to break up and any attempt to make it do so will cause considerable damage to structure.

Cultivating when soil is too dry is also to be avoided. Depending on the soil type, it can result in large, blocky lumps of soil that won't crumble naturally, also requiring more effort to cultivate, whether working by hand or machine.

Sandy loam soils are generally much more forgiving than clay loam soils. The window of opportunity for cultivating heavier soils is far narrower than for lighter soils, and the potential for causing damage by going a little too early or too late is far greater. When Biodynamic soil structure is well developed, the window becomes larger, and soil cultivation in general becomes much easier.

When digging by hand in the garden, we should use a garden fork rather than a spade. Why? When we use a spade, we create a series of cut surfaces, which smear over the air channels that allow the soil to breathe. These surfaces can dry out and become hard. Moreover, the cuts bear no relation to the natural rubbly structure of the soil; they arbitrarily cut the soil. Normally, when spades are used, blocks of soil are brought out and then bashed to get rid of lumps and reduce the soil to a fine texture. All structure is then lost. Soils treated this way often become compacted. On a larger scale, ploughing a soil that is too wet tends to cut the soil into blocks, the surfaces of each block being smeared over, sealing off the soil pores and stopping air penetration. These blocks become hard dry 'rocks'.

When digging with a garden fork, the tines do not create cut surfaces as a spade does. Thin tines are best, not wide tines as in a spading fork. We lift out a spit of soil and tap it with the fork just enough to break it down into smaller, naturally broken pieces, retaining as much of the rubble structure as possible. If the soil is a light sandy loam we must be very careful, maybe just lightly touching it with the fork to break it up enough without losing the structure. If the soil is a heavier clay loam, we may have to hit it harder and perhaps more times to break it up sufficiently. The more developed the structure, the less we will have to work it. We must watch the soil carefully as we work, to ensure we retain as much of the open rubble structure as possible, and not let our minds drift off.

We should treat the soil according to the result we want. For instance, if we have chopped up a green manure crop and wish to incorporate it into the soil, we may roughly turn the cut material under and leave it to rest for a week or more before further working. If we want to sow large seeds, we can afford to leave the soil a little rougher. If we want to sow finer seeds such as carrots, we need to finish the top 40mm or so more finely. If we were to try to make the soil very fine to depth, we would lose too much structure.

A Note on Soil Inversion

It is often believed that inverting the soil is bad for the soil life. This may be so in a soil that is compacted or has been smothered by overly heavy mulching that has cut off air access to the soil. However, in a well-structured soil with many air pockets, as long as subsoil is not brought into the top layers, inversion does no harm whatsoever. This is borne out by years of experience of Biodynamic farmers and gardeners, not to mention the long history of the caring use of traditional mouldboard ploughs,

pulled by cattle or horses, which maintained soils of immense fertility and structure for over 1,000 years in Europe and elsewhere. Unlike modern mouldboard ploughs, these traditional mouldboards had a downward-pointing tip that tended to lift and break the soil and ameliorate the tendency of the mouldboard blade to cut a flat 'table top' under the surface, leading to the development of a hardpan.

Tractor Cultivation

Biodynamic growers aim to apply all the above principles of hand digging to tractor cultivation. Obviously, much more damage to structure can occur with large-scale machinery if we do not take care.

Size of tractor

Generally, to minimise soil compaction, we use the smallest tractor that is practical for our farm situation. But if we need to cultivate a thousand acres in a short window we will need a large-horsepower tractor. Dual back wheels help, though bear in mind Podolinsky's caution that, no matter the pounds per square inch (which can be reduced by, for instance, the use of caterpillar tracks), the total weight of the tractor still bears as a whole on the subsoil. He asked, 'Would you drive a 10 tonne vehicle, even if it had caterpillar tracks or balloon tyres, over a 5 tonne capacity bridge?'

Speed of cultivation

Soil should be worked at a moderate to slow pace, generally at walking pace. Soil should not be 'chucked' or thrown about, nor should dust be created.

Implements

Tined implements are preferred: they are the large-scale equivalent of a garden fork and, if used when the soil moisture is right, at the right speed, will break the soil naturally (at natural joints and crumb structures) rather than cutting, smearing or bashing it to dust. However, a range of other implements can be used, provided extra care is taken.

Tined implements – rippers

Rippers penetrate deeply into the soil, relieving compaction when used at correct moisture levels.

Rehabilitator plough

In his *Biodynamic Agriculture, Introductory Lectures, Vol. 1*, published in 1985, Podolinsky included a drawing of the shape of an ideal deep ripper shank and described its lifting action on the soil.

Agrilatina, Italy – an 80-hectare Australian Demeter Biodynamic Method market garden – developed the first successful version of this plough in the early 2000s. Australian engineer Mike Fix (Fix Engineering) developed the Australian version, the Rehabilitator plough, in 2005. It features sharply angled, relatively thin (20mm) tines with a continuous curve and two rows of 'tusk' rollers following the tines. The sharply angled tines lift and break the soil upwards from the tip; by the time the tines reach the soil, it is already broken. The plough comes in sizes from 1.5m (three tines) to 4m (13 tines), and also as a lighter version with one bar, three tines and one set of tusk rollers. The first row of tusk rollers faces forward, 'digging' the soil, while the second row faces backwards, pushing crop residue or green

manure into the soil. They are reversible to suit differing uses. Attachments are available, including seed boxes and coulters (for pasture work). In some soils, farmers have found the Rehabilitator less suitable for relieving soil compaction under pasture, since the soil can become too soft and liable to pugging in wet conditions. A useful alternative for pasture work is the New Zealand-made James Aerator (see p. 119 (or similar). According to users, this fractures and aerates the soil to a depth of 25cm without unduly disturbing the surface or leading to pugging.

The Rehabilitator plough can easily penetrate 50cm with the tines set at the shallowest angle

Figure 16.1. and 16.2. Rehabilitator plough. Photo: Fix Engineering.

Figure 16.4. Agrowplow. Reproduced from Agrowplow (www.agrowplow.com.au) with kind permission.

Agrowplow

The Agrowplow lifts and shatters compacted soil and subsoil without inversion. Agrowplow make a range of attachments, including coulters to cut through trash or pasture, rollers that break down clods, hillers, mulch blades and more.

Yeomans plough

P.A. Yeomans, developer of the Keyline ecological farming method, and his son Allan, developed a very efficient deep ripper system to accompany this farming method. Stump jump versions are made for rocky ground. Today Yeomans manufacture a wide range of attachments, including coulters for pasture work, hillers, crumble rollers and weed knives.

Figure 16.3. James aerator. Photo: Franco Santucci.

(deeper when set at a steeper angle, but requiring more effort); it can lift the soil 150mm without bringing up clay or mixing soil layers. The ripper tips have a uniquely efficient design. The Rehabilitator can penetrate deeply in even the hardest soils and be drawn through the soil with relatively little effort. Most deep rippers require much more horsepower to achieve the same depth. They generally press forward with the tines, *forcing* the soil open, rather than gently lifting and crumbling it naturally as does the Rehabilitator. Fix Engineering also make a range of other implements, including mechanical weeders.

Figure 16.5. Yeomans plow shanks. Reproduced from Yeoman's Plow (www.yeomansplow.com.au) with kind permission.

Single-tine deep ripper

This is useful for smaller areas and smaller tractors and is often fitted with a pipe-laying attachment. It acts similarly to the above implements, relieving subsoil compaction, and is often used across the slope or along the contour, with the rip lines 1m or so apart. Dig after doing a few runs to see how far the cracking extends. This implement can be used in conjunction with a chisel plough or light-tined implement to crack the subsoil and bring a bit more depth into the cultivation.

Other tined implements

Chisel plough

Chisel ploughs gently break soil along natural lines without turning the soil over. They have a similar action to rippers, but in the top 30–40cm only. They can also be used when soil is semi-dry to create a 'chiselling' action (rapid back-and-forth tine movement) that cracks the subsoil, relieving compaction. They are a good primary cultivation tool. Chisel ploughs (like

Figure 16.7. Massey Ferguson stump jump tiller.

all tined implements) will block if there is too much vegetation, so normally they are used after grazing or mulching of the vegetation.

Light-tined implements

These include the Massey Ferguson tiller. They suit smaller properties and lower-horsepower tractors. They can penetrate to around 30cm deep, often requiring added weight to go in fully.

S-tines (curly tines, triple K)

These are finer, springier tines and are useful for secondary cultivation, to break the soil into a finer tilth, and are also for weeding.

Figure 16.6. John Berends stump jump chisel plough. Reproduced from John Berends Implements (www. johnberendsimplements.com.au) with kind permission.

Figure 16.8. S-tines

Figure 16.9.-Scarifiers

Scarifiers

Scarifiers are often used on grain properties to prepare for sowing, either alone (especially following a previous season's grain crop) or in conjunction with a chisel plough or discs.

Other cultivation implements

Disc plough, disc harrows

These implements are not as gentle on the soil as tined implements and have a cutting action. However, used slowly and carefully under suitable conditions, they can be used effectively. They can also be used to roughly work in a heavy green

Figure 16.10. Small set of scalloped discs used to work in a green manure that has first been mulched.

Figure 16.11. The discs have done little damage, retaining much of the rubble structure after working in the green manure, going one way and then across. Tines will follow.

manure crop where a mulcher is not available to chop it up finely first, or after mulching.

Mouldboard plough

Mouldboard ploughs are not often used on Biodynamic farms, since they cut a 'table top' under the blades, which can lead to compaction (hardpan) unless followed by deeper-penetrating tines. However, the plough can be retrofitted with a downward-pointing tip on each blade to produce more of a lifting action ahead of the mouldboard, which is much better. Some Biodynamic farmers have used mouldboards to bury a pasture with too much bent grass, going into winter, and found that this effectively kills the bent grass during the winter months.

Rotary hoe

Rotary hoes have the potential to cause considerable damage to soil structure. If used at all, they should be operated at the slowest-workable PTO speed. Even then, they pulverise

Figure 16.12. Rotary hoe.

soil, cut a table top under the cultivation and can set back soil development for years. However, they have one very good use on a Biodynamic farm: chip hoeing. Chip hoeing entails setting the blades to just run under the surface of a pasture when the soil is semi-dry and coming into a warm, dry period. Leave for a week or two for the grasses to die, then work with a tined implement. This breaks up the flat bottom cut left by the rotary hoe blades, breaks up the soil, works the dead grasses in and begins the process of drying out any grasses – such as bent, couch, kikuyu – that would otherwise regrow.

The above is not an exhaustive list of possible cultivation implements. Others include no-till or light-till seeding equipment such as the culti-trash (small discs or tines) and no-till disc drill. Other implements can certainly be considered and tried, and advice sought from experienced Biodynamic farmers.

Cultivation Techniques

Learning to cultivate soil caringly and effectively is a key developmental process in Biodynamics. Although general pointers can be given,

every farm, every soil and every climate are different. Even on the same farm, different seasons can require different approaches, as will circumstances such as whether we are cultivating recently cropped land or established pasture.

As an example, the following describes how established pasture can be prepared for growing vegetables:

- During a warm, dryish period, when there is enough soil moisture but not too much, graze the pasture or chop it up with a PTO mulcher. The latter results in more organic matter being available to build humus and feed a subsequent crop.
- Chip hoe with a rotary hoe, cutting just below the crown of the grasses. Leave to rest for a few weeks to allow the grasses to largely die. (Tines or discs could be used instead.)
- Chisel plough one way and then straight away across at right angles. If there is some compaction, a deep ripper can be used – or, for instance, an Agrowplow one way, followed by a chisel plough at right angles. A Rehabilitator plough, if available, would be ideal, working the soil one way, then across at right angles.
- Leave to mellow for a few weeks.
- Cultivate with S-tines. This could be done a few times at intervals of a few weeks, depending how well the grasses have been digested and whether runners of kikuyu, bent or couch have dried out and died. It will also kill any newly germinated grasses or weeds.
- For a light-feeding crop, beds can be formed, weeds or grasses allowed to germinate

(this can be done a few times), followed by mechanical or flame weed control, then the crop can be sowed.

- If the soil is unlikely to support a heavier-feeding crop straight away, a mixed green manure can be sown, fed with organically allowable fertilisers if necessary, before sowing or transplanting a crop. If any raw manures are applied to the first green manure, a second, unmanured green manure crop is required before a food crop is grown, to ensure that all manure is transformed into humus and to eliminate any risk of *Escherichia coli* contamination of the vegetables.

This is just one example. So many different combinations are possible and all can be considered. The challenge is to find the best procedure for each particular farm and situation while retaining as much of the open, rubbly soil structure as possible. If one bears in mind the earlier advice on hand digging, and remains observant, experience will gradually produce the best possible outcomes. In Australia, the BDAAA assists newcomers with a mentor system, pairing farmers with someone with a similar farming enterprise, usually in the same district. The mentor can save new Biodynamic farmers years with timely advice until they find their feet. Similar mentoring occurs in Europe and Asia where the Australian Demeter Biodynamic method is established.

Chapter 17

Sheet Composting

Biodynamic sheet composting is an important process fostering the development of soil fertility and structure. It is a form of in situ composting which encompasses:

- The dynamic management of pasture in tandem with progressively deeper penetration of roots, humus development and soil structuring, particularly fostered by the use of prepared 500.
- The depositing of animal manures by grazing stock, optimised, in the case of cattle, by harrowing the manure after grazing.
- The 'topping' of pastures that have got too far ahead of the stock, as can happen in spring or other warm, wet periods.
- The mowing of non-grazed pasture in orchards, vineyards, olive groves, berry production and similar enterprises, and also the cultivation of the under-row or mid-row, with or without green manuring.
- The growing and working in of green manure crops, often with the application of animal manures before sowing.

Sheet Composting of Pasture

Grazing

Dynamic rotational grazing has been an integral part of the Australian Demeter Biodynamic agricultural method since the early 1950s, when many government agriculture departments were recommending 'set-stocking' whereby stock were permanently pastured on one paddock, with no rotation, and many decades before 'cell grazing' became popular.

For truly dynamic pasture management, many paddocks are required. Dairy farms often have 40–70 separate paddocks and subdivide these with electric wires. After a section has been grazed, the wire is moved to allow access to a fresh strip of grass. If the animals start re-grazing the already grazed sections, they have been there too long. If necessary, another wire can be set up to stop 'back-grazing' of the already grazed sections, allowing them to begin regrowing without further disturbance. Strip grazing is most applicable to cattle, which are easily managed with a single electric wire. Strip grazing can potentially almost double the productivity of an area. Poultry and pigs can be confined with moveable electric netting, which can be relatively easily moved to fresh pasture. Even on a small property, at least five paddocks are required to allow dynamic rotational grazing.

When livestock graze a section, a corresponding portion of old roots die back and become available to worms and microbes, which digest them and convert them into humus, at the same time creating new air channels. New roots then grow into the air channels, extending deeper and building soil structure with each new grazing. Under Biodynamic grazing management, roots can extend 2–3m into the soil, building structure as they go, in partnership with the biological activity fostered by prepared 500.

Pasture should be grazed at the optimum stage for the time of year and the type of stock. For instance, on dairy farms, depending on the season, rye grass should ideally be at the three-leaf stage, but in spring, when the grass is growing rapidly, it could have five or six leaves. Optimum height after grazing is 4–8cm.

In spring or at other times, grass may get too far ahead of the stock and be at risk of stagnating. In this case, the pasture can be 'topped', cutting it back a bit. This restores the pasture to dynamic growth and also provides additional nutrients as the cut grass is broken down and incorporated into the soil humus. It is better to top pasture with sharp blades than to 'tear' it as occurs with, for instance, a mulcher.

Pasture that is allowed to stagnate sees reduced root activity and consequent reduced structuring of soil. The exception is haymaking: it is very beneficial to allow pasture to fully express itself in spring for haymaking every few years.

Harrowing

When stock graze, they deposit their valuable manure and urine evenly around a paddock. Worms, microbes and other soil life set to work digesting the manure and urine, incorporating it into the soil colloidal humus. Sheep and goat manure is relatively small and more easily processed. Cow pats, however, are much larger and take longer to process. If left where they are deposited, they concentrate the valuable elements in one small area, in a soluble form pending their incorporation into the soil humus. Grass is oversupplied with soluble elements and becomes overblown and darker green. Cows avoid it for some time, thus wasting grass.

Biodynamic farmers harrow cow manure after grazing. This spreads the manure thinly over a wide area. It allows the manure to be more quickly incorporated into the soil humus and also spreads the valuable cow fertility broadly over the paddock, avoiding the over-concentration of elements in discrete areas and the wastage of pasture that results. Harrowing can be done at any time of the year except when

Figure 17.1. Dark green, overblown grass where cow pats were deposited and not harrowed.

the soil is so wet that tractor tyres leave an imprint. In warmer seasons in many areas, dung beetles become active, breaking down the cow pats so quickly that harrowing becomes a waste of time and fuel.

Pasture harrows come in various forms, including split truck tyres ('tyre smudgers'), which are best suited to undulating country because they 'flow' over undulations; rigid frame tyre harrows, which may be used on flatter ground; and various types of chain harrows, some with small spikes. Harrows are towed through gateways by attaching a chain to the end. Caution should be exercised where particular undesirable pasture species such as kikuyu and couch grass are present, since harrows can potentially collect runners and spread them about.

Non-grazed pasture

Pasture in vineyards, orchards, berry plantings, etc. is sheet composted differently. It is not generally desirable to introduce grazing animals in these horticultural operations, since this tends to interfere with the healthy cycling of nutrients. However, the same principle of maintaining dynamic pasture activity applies. In this case, the equivalent of grazing is slashing or mowing. Slashing has the same effect as grazing: after slashing, a corresponding amount of old roots die, and are digested by worms, microbes and other soil life. The cut grass left on the surface is also digested. As the grass regrows, new roots follow the worm channels down, penetrating progressively deeper. If grass is allowed to stagnate, this active cycle ceases. If the grass is cut too frequently, before it has had a chance to

Figure 17.2. Split truck tyre pasture harrows. Special equipment is required to split steel-belted tyres. Textile-belted truck tyres are more easily split, if obtainable. The bottom of the tyre bead is best removed to prevent build-up of soil and manure.

Figure 17.3. Paddock harrowed after the cows have moved on.

Figure 17.4. Rigid set of pasture harrows suited to flat land, Nathalia, Victoria. Photo: Mark Peterson.

manure should be sown without addition of any more manure, after working in the first, before growing any commercial crop. Prepared 500 is sprayed at or soon after sowing and when the crops are worked into the soil. This creates a deep sheet composting activity – the roots and digestive activity structuring and building humus from the surface right down to considerable depth. Green manuring is covered in detail in Chapter 18.

develop sufficiently, the effect will be similar to overgrazing, with root development regressing and becoming shallower.

Cultivation of the under-row or mid-row areas of a vineyard or orchard, with or without the growing of green manures, is also part of the sheet composting process.

Sheet Composting with Green Manures

Green manuring plays a valuable role in many farming enterprises including market gardening, fruit and nut growing, and cropping. It may also be used before re-sowing pastures in some situations. Raw manures, allowable non-water-soluble fertilisers, rock dusts or trace elements that are lacking may be spread before you sow the green manure. When it reaches an optimum stage (often as flowering starts), it is mulched finely and worked into the soil. If raw manures have been applied, a second green

Chapter 18
Green Manuring

Figure 18.1. Green manure crop.

In Biodynamics, plants are fed from the soil colloidal humus, as directed by the sun, rather than indiscriminately by water-soluble elements dissolved in the soil water. Colloidal humus is created by the soil biology (or in a composting process), particularly worms and microbes, and the process is greatly enhanced by the correct application of good-quality 500. Green manuring is an important method of building soil humus levels sufficient to feed subsequent crops.

A green manure is any crop that is sown or allowed to grow naturally and later incorporated in the soil, usually when it reaches optimum development, preferably before it becomes over-mature and woody. The green manure breaks down, with the assistance of prepared 500, and is transformed into colloidal humus, providing nutrition for subsequent food crops. Green manures can be used in many farming situations, including, but not limited to, market gardening, the production of fruit, berries and grapes, and grain growing. In vegetable growing, when colloidal humus levels become depleted, they can be replenished through compost application, but in most situations, after two (or occasionally three) crops have been grown in succession (depending on the quality of the soil and types of crops), the land is rested and a green manure crop is grown to restore humus and nutrient levels for subsequent crops.

Green manures can be even more beneficial than compost application, since they involve strongly growing plants with active roots, which structure soil and develop humus more deeply and effectively than compost application on its own. Subsequent crops then send their roots deeply (structuring the soil as they develop)

rather than feeding more shallowly as in the case of compost application.

What to Grow

Many plant species are suitable for use as green manures, and the more species grown together, the better. Legumes and non-legumes, deep- and shallow-rooting, each species brings something different to the soil. The 80-hectare market garden Agrilatina, near Rome, which uses the Australian Demeter Biodynamic method, guided by Alex Podolinsky, has up to 30 species in its winter green manure crops and up to 90 in its summer green manures.

It is important to let many plant families work together in a green manure. The balance of the two main families – legumes (Leguminosae, for nitrogen-fixing) and grains (Graminaceae) – together with other plant families, will vary according to the time of year, the purpose of the green manure, how long you can have the green manure in the field, the availability of the seeds and their cost. The exact composition is not as important as how the mixture works together as a unique complex. Each different plant species contributes a different quality to the soil as it grows and as it decays. Each species also interacts with its environment in a unique way, accumulating elements in varying ways, some specialising in particular trace elements absorbed from the air, brought up from deep in the soil, concentrated or transmuted (see pp. 25–27) in harmonious co-working with the teeming soil life.

Podolinsky pointed to the limitations of chemistry: for instance, the phosphorus content of a plant as determined by chemical analysis assumes the phosphorus *quality* is the same whatever the plant. He suggested that this is akin to maintaining that the same note played by all instruments in an orchestra is qualitatively identical, which is clearly not the case.

Finding the best mixtures of species for individual situations and seasons requires much experience and experimentation. As a general rule, the legumes should predominate, with around 70% legumes to 30% grains, adjusting to allow the inclusion of a wider range of species.

How Much to Sow

Seed mixes are generally spread at 20–40kg per acre (50–100kg per hectare). Depending on a number of factors, including the level of soil fertility, the higher rate can result in the overcrowding of plants. This can manifest in plants being forced upwards with no room for lively Biodynamic plant expression, the stifling and yellowing of lower leaves, and roots less able to develop freely and actively structure the soil. Biodynamic market gardener Darren Aitken has found that, sown at 20kg per acre (50kg per hectare), plants develop more refined and active upper plant expression and more refined root structure and, surprisingly, can produce a greater bulk of organic matter than can the crowded plants from a 40kg per acre (100kg per hectare) sowing.

Suitable Species

The following are some of the main varieties grown – many other species are suitable.

Legumes

- *Warm season:* beans, chickpeas, clovers, cow peas, dun/field peas, fenugreek, lucerne, lupins, mungbeans, vetch.
- *Cool season:* broad/faba/tic beans, fenugreek, dun/field peas, lupins, sub clover, vetch.
- *Tropical/subtropical:* cowpea, fenugreek, lablab, lupins, mungbean, soybean and many other beans, pigeon pea, Pinto's peanut, vetch, Wynn's cassia.

Non-legumes

- *Warm season:* barley, buckwheat, corn, Italian ryegrass, French and African marigolds (*Tagetes* spp.]), millet, mustard, oats, sorghum, sunflowers.
- *Cool season:* barley, brassicas (e.g. 'BQ mulch' – *Brassica napus/Brassica campestris*), Italian ryegrass, mustard, oats, rape (make sure it is non-GM), rye-corn, wheat.
- *Tropical/subtropical*: beans, buckwheat, French and African marigolds (*Tagetes* spp.), millet, oats, mustard, sorghum.

Inoculation of Legumes

Legumes are referred to as 'nitrogen-fixers'. Rhizobia bacteria form nodules on the roots of legumes; these bacteria absorb nitrogen (N_2) from the air and convert it into ammonium (NH_4), a form of nitrogen that can be used by plants and incorporated into soil humus. Specific species of rhizobia work with different plant species. In soils lacking biological diversity and richness, legume seed can be inoculated with the appropriate rhizobium to ensure optimum nitrogen fixation. Some seed companies supply the appropriate rhizobium along with the seed. In biologically active Biodynamic soils, this may not be necessary.

Bio-fumigation

Where root knot nematodes or root-rotting fungal pathogens are a problem, green manures can usefully include a high proportion of bio-fumigant species such as French and African marigolds and brassicas such as mustard and 'BQ mulch'. If a serious problem exists, a 100% bio-fumigant mulch can be sown, though there will then be no nitrogen fixation.

Feeding the Green Manure Crop

In highly developed, fertile soils, particularly where Biodynamics has been practised for some years, it is possible to simply allow whatever is growing to develop fully, without allowing it to go stalky and dry out, and work it in as a green manure without any fertiliser inputs. Some market gardeners find that this provides sufficient elements in the resulting colloidal humus to produce one good heavy feeding crop, followed by a light feeding crop.

However, in most situations, it is recommended to apply organically allowable inputs tailored to the individual soil and food crops intended to be grown, before sowing the green manure. There is no set prescription here, since every soil and climate are different. Experience must be the guide. As green manures are not intended to be food, inputs with a degree of water solubility, such as raw manures, are acceptable. In this case, two green manures in a

row are required as explained on p. 127.

Suitable inputs include anything allowed under organic standards, such as cow manure, poultry manure (fresh or pelletised), blood and bone, seaweed, lime, dolomite, gypsum, reactive rock phosphate, basalt rock dust, rock dust blends, guano and trace elements.

Here are some varied examples drawn from Biodynamic market gardeners:

- Cow manure 3 tonnes per acre (7.5 tonnes per hectare), blood and bone 150kg per acre (375kg per hectare), lime 250kg per acre (625kg per hectare), gypsum 250kg per acre (625kg per hectare), sulphate of potash 20kg per acre (50kg per hectare).
- Chicken manure $4m^3$ per acre ($10m^3$ per hectare), rock dust 1 tonne per acre (2.5 tonnes per hectare), lime 350kg per acre (875kg per hectare), gypsum 250kg per acre (625kg per hectare).
- Blood and bone 200kg per acre (500kg per hectare), rock dust 250kg per acre (625kg per hectare).
- Guano 50kg per acre (125kg per hectare), lime 200kg per acre (500kg per hectare).
- Guano 50kg per acre (125kg per hectare).
- No inputs at all. As mentioned above, on some properties, particularly where the soil is naturally of moderate to high fertility, where Biodynamics has been applied for many years and where enough area is available to allow sections to be cropped one year and rested the next, Biodynamic farmers are able to apply no inputs, allowing whatever species nature brings to grow, or sowing additional species during the resting

phase, and ploughing them in to prepare for the next season's crops. This works especially well where crops are well spaced and not grown too intensively – for example, just one row of brassicas per tractor width bed instead of two.

You can see how much variation there is between different properties. On very deep, fertile Dumbalk Valley soils in southern Victoria, I have grown very good green manures preceding market garden crops on just 50–60kg per acre (125–150kg per hectare) of guano and 200kg per acre (500kg per hectare) of lime. Poorer soils may need considerably more inputs, as in the examples above.

Fine Tuning

If irrigation is available, the crop can be particularly guided towards the deep development of roots. Podolinsky gave the following advice for green manures in Agrilatina's extensive greenhouses, but it can also be applied, with variations, in other situations. Sow the crop, then irrigate to 10cm depth. Spray prepared 500. When the plants are up, follow with another 10cm irrigation. When the plants are around 20cm high, water them deeply, ensuring that the soil is damp to depth (easier in deep sandy soil). Allow the soil to dry off from the top downwards. Prepared 500 stimulates the roots to follow the water downwards 'deeper and deeper, and the 500 activity therewith acts and structures deep into the subsoil. After four weeks, the soil can thus be penetrated to one meter.'[1]

One or Two Green Manures?

As green manure crops are not intended for consumption, raw and pelletised manures can be used. This is a form of sheet composting. Prepared 500 is sprayed around sowing time, after the inputs have been applied. This greatly accelerates the incorporation of the inputs into the soil humus.

If any raw manures have been used, a second green manure crop, without further manure inputs, must be sown after the first has been worked into the soil. This is to ensure that any water-soluble elements from the raw manures are fully incorporated into the soil humus and also to ensure that any pathogens in the raw manure, such as *E. coli,* are well and truly gone before food crops are grown. This is also a requirement under the Australian National Standard for Organic and Biodynamic Produce and in many other countries. Raw manures should never be used directly on food crops. People have died from E. coli infections resulting from the use of raw manure on food crops.

The second green manure can be sown as soon as the first has been chopped and roughly worked in and prepared 500 has been sprayed. It need not be grown to full size if time is pressing.

Topping

Sometimes the grains outgrow and dominate the legumes in a green manure. In such cases it can be worthwhile to 'top' the green manure once, cutting the top 150–300mm with a PTO mulcher, to allow the legumes to compete. Birthe Holt (in Denmark) also uses topping (to 20cm above the ground), slashing when the crop is flowering but it is too wet or too dry to work it in. This allows regrowth (and weeds to be smothered) before working the green manure in at a more appropriate time.

Working in the Green Manure

A green manure can be worked in at any stage of growth, but the best time is when there is a good bulk of material and before stems become too tough. The ideal stage for many crops is when flowering begins.

The ideal implement to use is a PTO mulcher, which chops the green manure up into small pieces that are relatively easy to work into the soil. Go fairly slowly to ensure a thorough job is done. The tractor wheels push some of the material down before mulching, and this material can't be properly chopped. Leave for a few hours to allow the pushed-down parts to stand up a bit, then go over again from the opposite direction.

When to plough in the chopped material after mulching the green manure depends on many factors: time of year, climatic conditions (temperature, wetness, windiness), soil type and condition, and so on. In some situations it is best to work the chopped green manure in straight after mulching. In other situations it is better to let it wilt for a day or two before working it in. Careful observation of the results over a period of years will refine the farmer's decisions. The chopped material is preferably worked into the soil with tined implements, though other options are discussed on pp. 118–122. Ideally, 10–30% of the material should still be showing after the first cultivation, and a second cultivation, a week or more later, will incorporate most of the remaining material. Prepared 500 should be sprayed just before or straight after the first cultivation to accelerate the conversion of organic matter into colloidal humus.

The best implement for working in a green manure is the Biodynamic Rehabilitator plough (see Chapter 16). The forward-sloping tines lift and loosen the soil, while the tusk rollers push the material in. Work the crop once and then repeat at right angles. This incoporates a large proportion of the material. The ground should then be left for a week to a few weeks, depending on the time of year, and a further working and another spray of prepared 500, if time allows, will then complete the incorporation of organic matter. When the organic matter has been built into colloidal humus by the soil biology (this can vary between 4 and 8 weeks, depending on many factors), food crops can be sown or transplanted.

Figure 18.2. First pass with mulcher, Dumbalk, Victoria.

Figure 18.3. Second pass in the opposite direction.

Figure 18.5. Working across at right angles to first working.

Figure 18.4. First working with discs.

Figure 18.6. Terradisc – one-pass incorporation of green manure. Photo: Mark Peterson.

There are alternatives to the Rehabilitator, though none is quite as effective. A chisel plough may be used, depending on the amount of organic matter and the soil type. In some situations, however, the tines may collect and drag material too much, despite it having been chopped by the mulcher. Weighted discs can be used slowly, once or twice in one direction, then once across at right angles. After a few

weeks of digestion, tines can be used to further incorporate the green manure. Podolinsky refers to this as 'stirring up the sheet composting heap'.

A mouldboard plough will turn most of the material under in one pass. However, this thorough burying can bring about an anaerobic situation resulting in an ineffective digestion process, sliming and waste of material and the leaching of nutrients. It can also create a 'table top' cut at the bottom of the mouldboards, further hampering the aerobic breakdown of material and leading to a hardpan. Follow-up with tines, after allowing time for some digestion, is essential in this case.

If a mulcher is not available, heavy discs can be used to cut and roughly work the crop in both ways, repeating after a few weeks if necessary and following up, after further mellowing, with tines. Or a rotary hoe can be used to chop up the crop before working it in. It's preferable not to use the rotary hoe to also work the crop in, since soil structure can be seriously affected.

An Austrian company, Pöttinger, makes a one-pass machine, the Terradisc, that is very effective at incorporating a green manure to a depth of 12cm in one pass. This is not so deep as to lead to anaerobic breakdown as can occur with a mouldboard plough. Later tine working will incorporate the partly broken down material more deeply.

Weeds as Green Manure

Barry Edwards, a third-generation Mallee (Western Victoria) farmer who started farming Biodynamically on his and his wife Fiona's 3,100 acres (1,250 hectares) in 1985, has pioneered the use of weeds as green manure. He noticed, over many years, that weeds tend to come to balance soil deficiencies and that, if these weeds are ploughed in at the appropriate time, they will redress the deficiency and no longer need to grow. For instance, heliotrope (potato weed) is very high in copper. Where soils are copper deficient, it grows 'prolifically over summer,

taking up copper with deep roots that penetrate the subsoil. If this plant is ploughed in as it starts to flower, or before it dries off, then your topsoil can be enriched with copper.' [2]

Barry and Fiona produce cereals and fat lambs. In the late winter or early spring before a cereal crop, Barry fallows with a one-way disc, ploughing in as much bulk as possible – in effect, green manuring clovers, grasses and weeds. He works the ground again with tines in early summer. Summer rains bring up summer weeds, which are left to grow. Conventional growers spray these to conserve moisture in the soil for the winter grain crop. Barry ploughs in the weeds as they begin to flower or as they start to dry off. He finds that this brings large nutrient gains and increased organic matter, which will be refined into colloidal humus by soil biology. Despite the transpiration of moisture caused by the summer weeds, the crops don't suffer from water stress, since the increased humus, holding additional elements, holds growing season rainfall more effectively.

Each different weed species gathers particular elements to balance particular soil deficiencies. Many of the weeds have deep taproots that bring up elements from deep in the subsoil. Some fix nitrogen in the soil. And all plants have the capacity to absorb nitrogen and other elements from the atmosphere. Provided that the plants are worked into the soil before they dry off (concentrating the elements in the seed or losing them back into the atmosphere), these elements can be incorporated in the soil humus, redressing deficiencies. Barry needs to apply only 3kg of phosphate per hectare to crops (as rock phosphate, two crops in five years),

whereas his conventional neighbours use 12–
20kg per hectare. Yet his yields are just as good.

Barry finds that with this system, allowing
the weeds to fully express themselves, the weeds
will not need to grow for some time. This results
in better-nourished crops and livestock (no stock
minerals have been needed for the sheep since
the mid-1990s) and much fewer weed problems
in crops. Barry's conventional neighbours
spray their barley grass and brome grass
with glyphosate every year and have ongoing
problems with them. Barry, using his system,
has no problem with barley grass or brome
grass. In 1992, he had a thick growth of soursob
(*Oxalis*). Soursob accumulates magnesium and
can be dangerous for sheep because it locks up
calcium in their bodies. Barry ploughed in the
lush soursob. This addressed the deficiency it
had come to balance; since then, soursob bulbs
have remained present, but the plants never
develop beyond 2 cm across. They are simply not
needed any more. Podolinsky suggested that this
redressing of imbalances by weeds accumulating
what is needed 'happens via the "sense" our
Biodynamic preparations activate'.

FOOTNOTES FOR CHAPTER 18
1.   Alex Podolinsky, *Biodynamics Agriculture of the Future*,
     BDAAA, Powelltown, Victoria, 2000, p. 39.
2.   Barry Edwards, 'Weeds, Just What Your Soil Needs', *Biody-
     namic Growing*, 9 (2009), p. 31.

Chapter 19

Working with Natural Rhythms

The world pulses with myriad rhythms and cycles – day, night, spring, summer, autumn, winter, dry season, wet season, moon cycles, tides and many more. We respond to and incorporate many of these rhythms in our daily lives. Biodynamic growers recognise and use a range of natural rhythms and cycles beyond those normally recognised, to great advantage. These include daily rhythms, the lunar cycle and the movement of the moon through the zodiacal constellations. The lunar and zodiacal rhythms have been the subject of extensive research by Biodynamic researchers such as Lily and Eugen Kolisko, Maria Thun and Agnes Fyfe.

Figure 19.1. Starry Night by Vincent van Gogh, 1889.

Daily Rhythms

From around 3am each morning, the earth's moisture belt begins to rise – sometimes evidenced by early morning mists – and begins to descend again (e.g. through dew formation) around 3pm Alex Podolinsky said that Van Gogh's beautiful *Starry Night*, which conveys so much of the richness of the heavenly bodies, depicts, among many other things, the earth's moisture belt beginning to descend. We use this rising and falling rhythm to guide many farm or garden activities:

- The soil spray, 500, is (with a few exceptions) sprayed after 3pm and up until 2am. During this time, the earth's 'drawing in' or 'breathing in' of the moisture belt helps draw 500 into the soil more effectively.
- Conversely, 501, the light spray, which works on the upper plant, is sprayed after sunrise, as the earth 'breathes out', until 10am (earlier in summer in light-filled climates such as Australia's).
- Seedlings are best transplanted after 3pm, since their roots take better then. The process can be aided by lifting them in the morning and keeping them in a shady place to slightly starve them before transplanting them in the afternoon. This does require some discretion and experience.

Moon Rhythms

We are all aware of the enormous influence the moon has on water, the tides being the most obvious aspect of this. Is it not possible that the moon also influences water in plants? Important moon cycles considered in Biodynamics include:

- Waxing and waning, full moon, new moon.
- Ascending and descending (moon becoming higher or lower in the sky).
- Apogee (moon farness), perigee (moon nearness).
- Nodes (when the moon's path crosses the sun's path, twice each month).

In the past, farmers and gardeners in many parts of the world used moon cycles for many activities, including sowing seeds. In Devonport, Tasmania, in the early 1970s, I met an old gardener who still sowed by the moon and told me his parents and grandparents had always done that. Moon sowing charts were commonly available in almanacs and Tasmanian newspapers at that time.

Rudolf Steiner suggested in his agricultural lecture series that the moon, planets and stars have a strong influence on plant growth. Much research has been carried out based on his suggestions. As Biodynamics has developed worldwide, lunar and astrological sowing calendars have come back into fashion and become widely available. Many, however, are based on untested theories or traditional astrology. Most bear no relation to the actual position of the traditional zodiacal constellations and have no basis in reality.

Biodynamics is science based. Lily and Eugen Kolisko carried out extensive experiments based on Steiner's suggestions for decades from 1920 onwards. Eugen died in 1939, but Lily continued her studies for decades afterwards.[1] Many aspects of the moon's influence on plant growth were investigated, and many of their findings have been successfully applied by Biodynamic farmers and gardeners. One of their key discoveries was that seed sown 2 days before full moon produced much better crops than seed sown 2 days before new moon. Following on from the Koliskos' work, Agnes Fyfe, too, carried out experiments on the moon's influence on plant growth.[2]

The Koliskos' results for sowing closer to full moon or new moon were consistent. Although there were unexplained variations, the results were always in favour of sowing 2 days before full moon. Sometimes a crop sown at a subsequent full moon was better than that sown at the previous full moon; sometimes leaf growth predominated, sometimes root growth; and so on.

It was German researcher Maria Thun who found an explanation. From the 1950s, she intensively studied the results of sowing a wide variety of crops according to various moon cycles. She discovered that, depending which zodiacal constellation was behind the moon at the time of sowing, the plant would be directed more towards root, leaf, flower or fruit.

She discovered that the constellations Taurus, Virgo and Capricorn influenced more towards root growth; Cancer, Scorpio and Pisces more towards leaf; Gemini, Libra and Aquarius more towards flower; and Aries, Leo and Sagittarius more towards fruit and seed. This, combined with sowing on a waxing moon (better towards full moon), would eliminate the variation in results. For instance, carrots sown at a root time would consistently develop larger roots and smaller leaves, whereas carrots sown at a leaf time would develop larger leaves and smaller roots. Lettuce sown at a flower or fruit time would be more likely to run prematurely to seed, rather than concentrate on leaf as when sown at a leaf time. In field trials in Australia, pumpkins sown at a fruit/seed time produced many times the weight of pumpkins compared with those sown at a leaf time.

Thun's observations were based on the actual position of the zodiacal constellations in relation to the earth and moon. This is called the 'sidereal zodiac', as opposed to the 'tropical zodiac', which is based on the positions of the constellations 2,000 years ago and thus does not reflect the present reality, since each year the constellations are slightly behind the position they were in a year previously ('precession of the equinoxes'). In the last 2,000 years, the constellations have moved backwards approximately 23 degrees. Practitioners of the Australian Demeter Biodynamic method use the sidereal (or 'actual') zodiac as used by Maria Thun. We recognise that the 12 zodiacal constellations do not each occupy an area of 30 degrees in the sky. Some are larger, thus taking longer for the moon to move across them, and some are smaller, thus taking less time for the moon to move across them. Virgo is the largest constellation, which takes more than twice as long for the moon to pass through as Libra, the smallest.

The Responsiveness of Plants Varies

It is very important to understand that plants grown within nature's organisation (see Chapter 1) are much more sensitive and responsive to moon and zodiac influences than plants grown outside nature's organisation, which are not functioning normally as plants. In most cases, it would be pointless to apply the following guidelines to plants artificially fed with water-soluble elements (whether derived from artificial fertilisers, raw manures or poorly made compost), since little or no result would be expected.

Let us summarise the findings of the Koliskos, Fyfe and Thun, together with observations and field trials carried out by Podolinsky and the farmers of the BDAAA.

Waxing, Waning, Full Moon, New Moon (29.5-day cycle)

- During the waxing moon, moisture content in the upper plant steadily increases, reaching a maximum at full moon. Moisture content then falls steadily, reaching a minimum at new moon.

- Seed is best sown during a waxing moon, preferably closer to full moon, but no later than 2 days before full moon. Seeds sown 2 days before full moon bring the best results. Seeds sown 2 days before new moon grow much less successfully, resulting in smaller crops, while sowing at the first and last quarters produces intermediate results. The Koliskos carried out comprehensive experiments from 1926, based on Steiner's suggestions, to establish this. They reported that wheat, barley and oats sown 2 days before full moon yielded 25% more than the same crops sown 2 days before new moon. For peas and beans there was a corresponding 80% increase.[3] For tomatoes

it was 60–100%. The waning moon is normally avoided for sowing, but third-quarter sowings may be needed when the best sowing times have been missed or when a regular supply of fresh seedlings is needed. Last-quarter sowings are likely to be less successful.

- As with other crops, pasture is sown during the waxing moon (at a leaf time – see p. 139). However, if a drought is anticipated, Podolinsky suggests that sowing towards new moon (and at a root time) will encourage root formation rather than leaf, ensuring that the pasture first establishes well underground. This can also apply to spring-sown pasture: abundant rain will encourage abundant leaf growth without adequate root foundation, leading to the plants drying up too quickly when the rain stops.

- Crops harvested at or towards full moon contain more moisture, will be juicier, look full and attractive and are excellent for eating fresh. However, they will not keep as well. Crops harvested at or towards new moon will have a lower moisture content; this is the best time to harvest crops that are to be stored or dried. Hay cut at or near new moon will dry more quickly; however, weather conditions and the degree of maturity of the grasses normally take priority.

- Pasture looks at its best at full moon, more vibrant and glowing, and will look considerably less so at new moon.

- Soil and root activity are more pronounced during a waning moon. Activities such as transplanting, working in green manure crops and making and spreading compost are best done during this time. Spraying 500 during a waning moon will be slightly more beneficial, though other factors such as weather conditions and soil moisture are more important considerations.

- Grafting is best done towards full moon. The scion will take better then owing to rising sap in the stock. Pruning is best done during the waning moon.
- Timber is best felled on a waning moon, even better at new moon. It will then dry more rapidly and with less cracking (something particularly important for woodworking purposes).
- Drenching (with natural, non-chemical drenches) of animals affected by worms is best done at full moon.
- Castration of animals is best done during a waning moon.

Ascending and Descending Moon (27.3-day cycle)

During the 27.3-day cycle, the moon is seen to climb steadily higher in the sky until it reaches its highest point (fully ascended) and then to steadily descend until it reaches its lowest point (fully descended).

An ascending moon has similar effects on plants to those of a waxing moon, with maximum effect at fully ascended. A descending moon has similar effects to a waning moon. Thus, for instance, an ascending moon is the best time for:

- Seed sowing.
- Grafting.
- 501 spraying.

A descending moon is the best time for:

- Activities related to soil/root/fermentation activity, such as 500 spraying, cultivation, composting and transplanting.
- Pruning.

Apogee/Perigee (27.6-day cycle)

The moon's path around the earth is an ellipse rather than a circle. Thus the moon is sometimes closer to earth and sometimes further away. Its closest point is called 'perigee', and its furthest point is called 'apogee' (A = away!).

When the moon is closest to earth, its effect on plant growth is magnified. When it is furthest away, its effect is minimised. When full moon, fully ascended and perigee coincide, the chances of fungal diseases and insect attack increase. Potatoes are best sown at apogee.

Putting It All Together

The various moon cycles operate independently of each other and have slightly different durations. This means that the ideal time to carry out an activity according to one cycle will not necessarily coincide with that for another cycle. When favourable times happen to coincide, the effect will be magnified. However, the cycles often clash.

What to do? For seed sowing, pruning, harvesting, grafting and timber felling, the waxing/waning cycle should take precedence. For activities related to soil, root and fermentation, and 500 and 501 spraying, the ascending and descending cycle should take precedence.

Moon and Zodiacal Constellations

Seed sown when a zodiacal constellation is behind the moon (as seen from the earth) at a particular time will be influenced as noted on p. 139. This applies to sowing seed and can also influence other activities:

- Thun suggested that weed germination could be encouraged by cultivating with Leo behind the moon, and that the resulting weeds can be more effectively destroyed by cultivating 11 days later, when Capricorn is behind the moon.
- Preparation 501 can influence the plant more towards root, leaf, flower or fruit, depending which constellation is behind the moon at the time of spraying. However, it should not be used to deliberately push a plant too much in one direction; sowing the seed at the correct time is enough. Recording the time when 501 is sprayed can aid one's observation of the plant's subsequent growth trends.

As mentioned on pp. 139–140, these zodiac influences will only be observed in plants growing within nature's organisation. Field trials carried out in Australia by BDAAA members resulted in a four-fold increase in the weight of pumpkins sown at a fruit time compared with a similar acreage sown at a leaf time.

Nodes

A further complication arises with the fact that the moon and sun paths cross twice each month. This crossing point is called a 'node'. A node that occurs when the moon is ascending is called an 'ascending node', whereas a node that occurs while the moon is descending is called a 'descending node'. Podolinsky describes how, on node days (specifically from 24 hours before until 24 hours after the actual node time), there may be disturbances that can adversely affect people's mood, cause more traffic accidents and adversely affect the mating of cows and the sowing of seeds.[4] A Victorian veterinary surgeon involved in bovine embryo transplants compared his records of egg fertilisation and the long-term results with the sowing calendar prepared by the BDAAA. He found that most fertilisations carried out at node times didn't result in successful pregnancies. Of those that did, most resulted in bull calves. The heifers (females) that were born often had bad temperaments, which made them troublesome milkers. As a young married man, I asked Podolinsky whether this might also apply to humans; he reassured me that humans are much less affected by such influences than animals!

For seed sowing, we need to ensure that the important 36–48-hour period of seed development (see pp. 142–143) does not overlap the 24-hour period either side of a node. That is, we do not start sowing until at least 24 hours after a node time and do not sow later than 60 hours (preferably 64) before a node time.

Actual Timing of Sowing

It is most important to understand that the seed takes in the influence of the constellations (mediated by the moon) at a particular time in its development. As Podolinsky has demonstrated, seeds should be sown so as to have as much time as possible in the required zodiac sign, preferably at least 36 hours; 48

hours would be even better.[5] Given the varying durations of the various constellations, this can be quite a challenge. For the shorter duration constellations such as Libra (~35 hours), or when a sign starts at an inconvenient time such as 2am, or when a node is approaching, one can sow up to 10 hours before the start of the sign, to ensure that the 36-hour point after sowing is reached while the required sign is still in force.

Examples

1. During a waxing moon, the moon enters Sagittarius (fruit/seed) at 7am on 22 July, changing to Capricorn (root) at 8am on 24 July. Full moon occurs at 12.36pm on 24 July (see Figure 19.2). I want to sow peas (southern hemisphere). I would sow them from 7am (or earlier in the morning if desired) until around 1pm on 22 July. This gives at least 43 hours under Sagittarius. Being 2 days before full moon also makes this a perfect time to sow.

JULY 22	JULY 23	JULY 24
Sagittarius 7am **FRUIT/SEED**	Fully ascended 1.10am	O Full moon 12.36pm Capricorn 8am **ROOT**

Figure 19.2.

JANUARY 10	JANUARY 11	JANUARY 12	JANUARY 13
First quarter 4.12am Aries Midnight **FRUIT/SEED**		Taurus Midnight **ROOT**	Descending node 2.20pm

Figure 19.3.

2. During a waxing moon, the moon enters Aries (fruit/seed) at midnight on 10 January and moves into Taurus (root) at midnight on 12 January. A node occurs on 13 January at 2.20pm (see Figure 19.3). I want to sow beans (southern hemisphere). I would sow them just before dark, say 7pm on 10 January (5 hours before the start of Aries). This gives the seed 53 hours before the start of the next sign, Taurus, and, most importantly, 43 hours before the 24-hour node interference period starts at 2.20pm on 12 January (67 hours before the node). If I waited until 8am on 11 January, the seed would still have 40 hours before Taurus began (which would normally be fine) but only 30 hours before the node started to interfere (54 hours before the actual node) at 2.20pm on 12 January, which is not enough.

Nature Takes Precedence

In farming or gardening, we must keep a clear overview and not become too focused on following sowing charts or moon cycles at the expense of practicality. For instance, hay should be cut when the grasses have reached the ideal level of growth, balancing maximum nutrient levels with bulk. The weather is then the main controlling factor. Hay should be cut when a good window (several days with no rain) in the weather occurs around about the ideal maturity time. To wait for a new moon could mean weeks of delay if another window doesn't present.

It is not always possible to sow at the right time. Indeed, during some waxing moons, there will be no opportunity to sow one type of crop (root, leaf, flower or fruit) owing to a node or other factor. Another time could then be chosen, or you can wait until 2 days before full moon. Even if one has to sow during a waning moon, this is preferable to not sowing at all.

Planetary Influences

The planets influence events both within our solar system and on earth:

- The Koliskos reference many studies[6] that reveal planetary influences on sunspots, which in turn affect many aspects of life on earth, including climate.
- Podolinsky suggests that the sun, moon and closer planets (Mercury, Venus, Mars, Jupiter and Saturn) are each closely connected with specific groups of plants and have considerable influence over them.[7]
- Georg Schmidt ran trials in the 1970s in which he sowed trees, believed to be influenced by particular planets, when the relevant planet was in opposition to the sun (sun on one side of earth, planet on the other, at 180 degrees to each other). He found much better germination, root development, growth rate and resistance to disease and adverse climatic conditions when trees were sown at these times.
- Maria Thun found that, when planets are in opposition to other planets, the sun or the moon, beneficial effects can be found in plant growth. For instance, when seeds are sown at moon opposition Saturn, germination is very good and strong seedlings result, resistant to fungal diseases and insect attack. New Zealand Biodynamic

farmers have found that 501 sprayed at moon opposition Saturn has a particularly strong and long-lasting effect.

There is no doubt much still to learn about all this, and much research to be done. Some planetary influences have been strongly supported by experiments and field trials. The most relevant and practical of these for farmers is that rabbits and other vertebrates can be strongly deterred or entirely eradicated on a property by 'peppering' when Venus is in Scorpio. Peppering of rabbits involves burning the dried skin, when Venus is in Scorpio, on a very hot fire and finely grinding the resultant ash and charcoal. This ash is then distributed around the area to be protected – usually the boundary only, provided that rabbits inside the boundary are first dealt with by other means. Peppering is also used against insect pests and weeds. See Chapter 22 for more detail.

FOOTNOTES FOR CHAPTER 19
1. For example, Lily Kolisko, *Workings of the Stars in Earthly Substances*, Orient-Occident, Stuttgart, 1928; Lily Kolisko, *The Moon and the Growth of Plants*, Anthroposophical Agricultural Foundation, Bray, 1936; Kolisko and Kolisko, *Agriculture of Tomorrow*.
2. Agnes Fyfe, *Moon and Plant, Capillary Dynamic Studies*, Society for Cancer Research, Arlsheim, 1975.
3. Kolisko and Kolisko, *Agriculture of Tomorrow*, pp. 7–13.
4. Podolinsky, *Biodnamic Agriculture Introductory Lectures Volume 1*, pp. 111–113.
5. Ibid., pp. 118–122; Alex Podolinsky, *Living Knowledge*, BDAAA, Powelltown, Victoria, 2002, pp. 24–30.
6. Kolisko and Kolisko, *Agriculture of Tomorrow*, pp. 4–6.
7. Podolinsky, *Biodnamic Agriculture Introductory Lectures Volume 1*, pp. 160–171.

Chapter 20

Plant Pests and Diseases

Conventional agronomists often insist that the world population cannot be fed by organic agriculture, since yields are lower when artificial fertilisers, chemical pesticides, fungicides and herbicides are not used. This view is not supported by the evidence.

Among many studies, those run by the (US) Rodale Institute are well worth examining: it has run continuous comparison trials since 1981, comparing organic and conventional grain cropping systems in North America. They concluded that, after a 5-year transition period, the organic farms' yields were comparable to those of the conventional farms, that in drought times the organic farms produced up to 40% higher yields and that the organic farms earned at least three times more profit (up to six times in some years), used 45% less energy, were responsible for 40% less carbon emissions and leached no toxic chemicals into waterways.

In comparing food production on organic and conventional systems, the whole environmental picture should always be considered. Nutrient and chemical run-off from conventional farms has wide implications for food production. Consider the 'dead zone' in the Gulf of Mexico, an area of nearly 5,000 square miles (13,000km^2) where no fish can live. The cause? Nutrient run-off, primarily nitrogen and phosphorus, from the Mississippi, mostly caused by artificial fertilisers used in agriculture (and added to by sewage and lawn fertilisers). How much food production is lost as a result? This effect is not just found in the Gulf of Mexico, but affects fish production in many other parts of the world.

The Australian Great Barrier Reef is under threat from multiple sources, including warming seas causing coral bleaching. One of the main threats is nutrient and chemical run-off from conventional farms (sugar cane production is a major contributor), causing toxic algal blooms that adversely affect fish and other marine life.

Also in Australia, nutrient run-off from conventional farms causes toxic blue-green algal blooms under certain environmental conditions, leading to vast kills of fish and other aquatic life in rivers and streams.

So, when comparing food production of different agricultural systems, we can't ignore the wider implications of each system.

Many conventional agronomists have an outdated understanding of how plants feed. Many still don't understand the role of colloidal humus, don't understand that it can be completely consumed by fine hair feeder roots and don't understand that plants fed within nature's organisation are, overall, less susceptible to attack by pests and diseases than those grown outside nature's organisation. Their conclusion that organic farms must produce less because their plants are unprotected by chemical sprays is not science based. Consider the following.

Alex Podolinsky reported tests[1] run in Germany by L. Furst in 1962 that compared three apple orchards: Biodynamic, organic and conventional. The Biodynamic orchard was a section of the organic orchard that had been run Biodynamically for 6 years. Each orchard ensured that the apple trees received 140 pounds per acre of nitrogen. The source of the nitrogen, expressed as pounds per acre, is shown in Table 20.1.

Furst carried out egg counts of aphids and red spider mites on equivalent amounts of wood in the three orchard areas (Table 20.2). The Biodynamic orchard, with very low egg counts, did not need to take any action. The organic orchard sprayed malathion (then considered to be a low-toxicity pesticide and quasi-organic, though it would not be allowed under any certification standards today) because their red spider mites were causing problems. The conventional orchard recorded very high egg counts and sprayed DDT and phosdrin to counter them.

The experience of Biodynamic (and good organic) farmers has been that healthy, biologically active, well-structured soils that are adequately supplied with humus produce strong and resilient plants that have relatively

Source of Nitrogen	Biodynamic	Organic	Conventional
Microbes and humus	60	60	25
Clovers	52	15	12
Added	28 (blood and bone)	65 (blood and bone)	105 (sulphate of ammonia)

Table 20.1.

Species	Biodynamic	Organic	Conventional
Aphid	1	1	400
Red spider	18	80	6200

Table 20.2.

few problems. As a BDRI certification inspector of over 20 years' standing, I have observed that many Demeter Biodynamic farmers, at their annual inspections, report having needed to use no substances whatsoever for pest or disease control in the previous 12 months.

Plant Pests

Why do insect pests attack plants? Most insect 'pests' actually play a valuable role in balancing or beneficially influencing nature, whether we understand it or not. Many types of insect pests have a specific role in removing or attacking plants that are unbalanced – for instance, plants that are:

- Grown outside nature's organisation and therefore not functioning naturally because of an oversupply of soluble elements.
- Affected by too much water owing to

excessive rainfall or irrigation, poor drainage, or overcrowding or shading.

- Weak, whether because soils lack sufficient elements; because soils in the early stages of Biodynamic conversion have not yet reached the stage where they can support plants of high vitality; because of environmental factors such as drought, wind, sudden cold snaps or hail; because of planting too early or too late; or because of other factors.

Examples of these types of pest include: caterpillars such as tomato budworm, corn earworm and armyworm; sucking insects such as aphids, whitefly and red spider mite; and chewing pests such as lucerne flea, slugs and snails. These pests usually present little or no problem to Biodynamic farmers (after 3–5 years of soil development), who use Biodynamic preparations and other techniques to develop soil health and to balance the factors that encourage their presence.

Other types of insect pests are not directly linked to the health or otherwise of the plant. For example codling moths, fruit flies and white cabbage butterfly caterpillars. Such insects, though seen as pests, have wider roles such as pollination, and providing food for other insects and for bats and birds. Every living thing has its place and its right to exist in the great symphonic harmony that is nature. For instance, it has been suggested that codling moths have a natural role in helping thin heavy apple and pear crops, since 'stung' fruit soon fall from the tree. After 50 years of Biodynamics on the family orchard at Merrigum, Victoria, Lynton Greenwood found that the trees had become

so healthy that a codling sting would heal over and the fruit would remain on the tree until harvest, when it would be useable for juicing. For those insect pests that do not come as a result of imbalance, we have many natural methods available to reduce their impact.

Firstly, let's look at the Biodynamic approach to building health and vitality, balancing environmental imbalances and dealing with pests that don't arise from imbalance.

Biodynamic techniques

These are covered in more detail in other sections of the book.

Soil development

The development of deep, well-structured, well-drained, humus-rich soils is the basis of plant health. Prepared 500 is the main developer of healthy soil. Raised beds or rows are used to improve drainage where necessary (see Chapter 6).

Preparation 501

This is used when excess water or insufficient light is present in the environment and plants, to bring more light into plants that would otherwise become attractive to many types of sap suckers and caterpillars (see Chapter 12).

Preparations 500 and 501

These two preparations can be sprayed together (after stirring for 1 hour) to quickly strengthen plants that are under attack. This technique has been used very successfully against red-legged earth mite and lucerne flea in crops and pastures (see pp. 87–88).

Valerian (507)

Valerian brings warmth and healing to plants that have been or are about to be affected by sudden cold snaps, or have been battered by storms (see pp. 6, 12–14).

Peppering/charcoaling

In his 1924 lecture series, Rudolf Steiner suggested that pest insects and their larvae, vertebrate animals and weeds could be countered by burning the insect, animal skin or weed seed at specific times and distributing the ashes over the area required. This is covered in detail in Chapter 22.

Casuarina tea

Casuarina tea is a valuable anti-fungal agent. When it is allowed to ferment, it also has strong insecticidal properties. See pp. 154–155 for details.

Companion planting

In his agricultural lecture series of 1924, Steiner talked of the need to find the 'necessary neighbour plants' for different crops. He mentioned the beneficial influence of cornflower, dead nettle and sainfoin on grain crops, horseradish on potatoes, yarrow on many crops and other positive or negative combinations. This has inspired Biodynamic practitioners, over the years, to carefully observe and record beneficial and antagonistic plant combinations. The expanding body of knowledge, incorporating many practices from traditional farming, became known as 'companion planting'. Some plants have repellent properties against pests of other plants. Many books are available on this topic.[2]

Biodynamic treatment of fruit trees

Biodynamic methods and materials have been developed to enhance the health of fruit trees and protect them from pests and diseases (see Chapter 23).

Natural plant sprays

Garlic

Garlic is useful against a range of pests, including cockchafers, wireworms, aphids, cabbage white butterflies and codling moth. Finely chop 120g of garlic and mix with 15ml of light, odourless mineral oil such as liquid paraffin (*not* kerosene). Mix 20ml of liquid potassium soap such as NatraSoap with 1 litre of hot water; alternatively, mix 10g of soft soap in 1 litre of hot water. Mix all the ingredients together well. Strain and bottle. Use 1 part to 99 parts of water, increasing the strength if necessary.

Stinging nettle

Stinging nettle (*Urtica dioica*) can be used against aphids (e.g. in vineyards) and also has a mild anti-mildew action. It is not very effective on mildew by itself, but, combined with copper or sulphur, it reduces the amount of copper or sulphur required. One kilogram of fresh nettles are added to 5 litres of cold water. Heat until the mixture boils, then take it off the heat and allow it to infuse for 20 minutes. Make up the volume of water to 40–50 litres and add 1% by weight of powdered clay (e.g. kaolin clay, obtainable from potters' suppliers); this amount can be sprayed over 1 hectare. Don't use clay together with copper.

Wormwood

Wormwood helps repel various pests including slugs, aphids (particularly on beans) and moths such as cabbage moth and codling moth. Place 400g of fresh wormwood in 5 litres of cold water. Bring to the boil and simmer over low heat for 5 minutes. Add 45 litres of water. This amount will cover 1 hectare.

Long-term seed development

During its lifetime, a living organism is exposed to many environmental influences, which can affect its phenotype (the physical properties of the organism). Until recently, it was believed that these phenotypical changes would not affect the organism's genotype and so could not be inherited. However, many experimental findings are challenging this assumption. For example, changes in skin colouring in a genetically identical colony of mice, during their lifetime, are passed on in the genes to their offspring.

It appears that the genotype is much more fluid than has previously been assumed. There are strong indications that the upright, vibrant plants produced by Biodynamic methods may be able to pass on this vibrancy to the next generation. Melbourne Biodynamic gardener and seed saver Rod Turner grew capsicums Biodynamically from conventional bought seedlings (a non-hybrid variety), saved the seeds and the next season grew the same variety of bought seedlings next to seedlings he sowed from the saved seed. Photos of the capsicum plants growing next to each other showed distinct changes in form: more uprightness and more harmonious leaf arrangement, typical of Biodynamics, were evident in the plants grown

from the saved seed. More research is needed to follow up this initial observation.

I have found that peas grown from conventional pea seed are quite susceptible to mildew in adverse conditions, but that, after I have grown the peas Biodynamically and saved the seed for three or four generations, they are much stronger and far less likely to be affected by mildew. Again this indicates the inheritance of acquired characteristics.

Plants that are exposed to insect attack often produce substances that help the plant resist the insect in question. Could it be that these phenotypical changes can be passed on to the next generation via the genes? A long-time Biodynamic gardener from Geelong, Victoria, George Abbott, carefully saved seed from several species of brassicas for decades, with appropriate isolation before the first flowers opened, to avoid crossing between species. After many years, he found that the caterpillars of white cabbage butterflies no longer attacked his brassicas. It appears they had developed some degree of resistance to white cabbage caterpillars and passed those qualities on to subsequent generations, the effect increasing with each generation. This is an immensely promising field for further research and seed development and is why every country should endeavour to develop as wide a range of high-quality Biodynamic seed as possible.

Beneficial insects

Nature loves balance. Unbalanced plants are quickly attacked and broken down by insects, fungi, bacteria, earthworms, etc. and transformed into colloidal humus to feed other plants. Insect pests are generally kept in balance by other, predatory insects. Beneficial insects and arachnids include lacewings, ladybirds, small wasps, praying mantis, predatory mites and spiders.

Before resorting to (natural) sprays or dusts to kill pests that appear to be getting out of hand, beneficial insects should always be considered first if this is practicable. Whenever any spray, however natural, is used to kill pests, it is highly likely that beneficial insects, that are there naturally to deal with the pests, will also be killed. Beneficial insects work hand in hand with good general Biodynamic management.

It is always desirable to maintain a wide range of flowering plants to attract beneficial insects, whether it be in a good mixed grazing pasture, in pasture under an orchard or in a market garden. Some organic seed companies sell seed mixes of such plants. For instance, in Australia, Green Harvest sells a Good Bug Mix that includes lucerne, fennel, buckwheat, rue, white clover, yarrow, cosmos, Queen Anne's lace, sweet Alice, dill, tansy, Chinese mustard, parsley and daisies.

There is now a well-developed body of knowledge about beneficial insects that attack insect pests. In many countries there are reliable suppliers of these beneficials, such as Bugs for Bugs in Australia. For instance, lacewings and ladybirds, as eggs or mature beetles, are readily available in many countries as predators of aphids. *Trichogramma* wasps are used in Australia to deal with many moth species including *Heliothis*, loopers, codling moth and light brown apple moth. *Aphytis* wasps are useful in attacking various scale insects. These are but a few examples.

Figure 20.1. Ladybird on celery leaves.

Birds and bats

Birds and bats are wonderful, hard-working insect predators and their presence should always be encouraged. Every farm should develop extensive shelter belts of trees and shrubs, not only for the shelter they provide, but also as essential habitat for birds. Birds of prey, including owls, can be attracted by setting up nesting boxes for them. They greatly assist with rodent control. Bats can be encouraged by setting up special bat houses.

Other biological controls

Organic and Biodynamic farmers have a wide and growing repertoire of effective biological controls for pests. These include pheromone traps for insects such as codling moth and fruit fly, bacteria such as *Bacillus thuringiensis* for caterpillars, and viruses such as Madex virus for codling moth.

Cultural controls

These are interventions such as crop covers; fruit cages or bags that exclude insects such as fruit fly; sticky traps, such as the wide, yellow, sticky bands that are wrapped around fruit tree trunks to stop earwigs (they are put on early in the season, after 9am, when the earwigs have returned to the ground after feeding in the trees overnight, to prevent their return); cardboard or hessian that is wrapped around tree trunks to trap codling moth grubs; and other types of insect traps, with or without pheromone lures.

Integrated pest management

Integrated pest management (IPM) is a systematic approach to pest control that aims to first use natural means of controlling pests, including cultural controls and biological controls. It involves developing a detailed knowledge of the pest or pests, taking steps to prevent a problem developing, careful observation of the developing crop (including monitoring of traps), using mechanical and biological interventions when necessary, and, finally, using suitable sprays. Integrated pest management is a very good development that has encouraged conventional farmers to avoid as far as possible the use of chemical sprays. Of course, when IPM is used by organic or Biodynamic farmers, the sprays or dusts used will only be those allowed under national organic/Biodynamic standards.

Weevils in stored grains

Biodynamic and organic growers have several reliable methods of controlling weevils in stored grain:

- Aeration systems that draw cool air into the silos overnight. This keeps the grain cool, maintains viability and controls weevils.

- Carbon dioxide gas is introduced into sealed silos as required.
- A nitrogen generator (a compressor pushes air through a molecular filter that separates out nitrogen molecules) increases the nitrogen content of the air in the silo from 79% to 99% and reduces the oxygen content to less than 1%.
- Silos can be painted with a special white insulating paint. This keeps the silos 10°C cooler in summer, which is better for the grain and helps against weevils.
- Diatomaceous earth is added to grain as it is augered into the silo.

Organically allowable materials for pest control

These may vary slightly from country to country, but are generally very similar to allow international recognition of the equivalency of national certification standards in order to facilitate organic trade. In Australia, the National Standard for Organic and Biodynamic Produce lists all allowable organic inputs; the list of permitted materials for plant pest control appears in Figure 20.2 (overleaf).[4] In other countries, consult your own national organic standards.

Plant Diseases

Most plant diseases are caused by one or more of the following:

- Excess water in the plant, soil or wider environment, and/or poor drainage.
- Insufficient light.
- Lack of air circulation, excessive shading or overcrowding of plants.

National Standard for Organic and Bio-Dynamic Produce – Edition 3.8

Appendix C - Permitted materials for plant pest and disease control

Where wetting agents are required, caution needs to be exercised with commercial formulations as these may contain substances prohibited under this Standard. Acceptable wetting agents include some seaweed products, plant products (including oils) and natural soaps.

Table A2 Plant pest control

Substances	Specific conditions/restrictions
Ayurvedic preparations	None
Baits for fruit fly	Substances as required by regulation. Baits must be fully enclosed within traps.
Boric acid	None
Biological controls	Naturally occurring cultured organisms e.g. *Bacillus thuringiensis*.
Diatomaceous earth and naturally occurring chitin products	None
Essential oils, plant oils and extracts	None
Homeopathic preparations	None
Hydrogen Peroxide	None
Iron (III) phosphate	None
Light mineral oils, such as paraffin	None
Lime	None
Natural acids (e.g. vinegar)	None
Natural plant extracts excluding tobacco/	Obtained by infusion and made by the farmer without additional concentration
Pheromones	None
Potassium Bicarbonate	None
Potassium permanganate	None
Pyrethrum	Extracted from *Chrysanthemum cinerariaefolium*
Quassia	Extracted from *Quassia armara*
Ryania	Extracted from *Ryania speciosa*
Seaweed, seaweed meal, seaweed extracts	None
Sea salts and salty water	None
Sodium bicarbonate	None
Sterilised insect males	Need recognised by certification organisation where other controls are not available.
Stone meal	None
Vegetable oils	None

Figure 20.2. Appendix C, Table A2.
National Standard for Organic and Bio-Dynamic Produce, Edition 3.8: November 2022, Department of Agriculture, Fisheries and Forestry, Canberra. Reproduced here under a Creative Commons Attribution 4.0 International Licence. No changes have been made. Copyright is owned by the Commonwealth of Australia. Disclaimer: The Australian Government acting through the Department of Agriculture, Fisheries and Forestry has exercised due care and skill in preparing and compiling the information and data in this publication. Notwithstanding, the Department of Agriculture, Fisheries and Forestry, its employees and advisers disclaim all liability, including liability for negligence and for any loss, damage, injury, expense or cost incurred by any person as a result of accessing, using or relying on any of the information or data in this publication to the maximum extent permitted by law.

- Unassimilated excess elements in plant tissues, caused by soluble elements dissolved in the soil water (feeding outside nature's organisation), whether from water-soluble fertilisers, raw or partly decomposed manures or poorly made composts.
- Weakness caused by insufficient elements, drought or other environmental factors.
- Damage caused by wind, hail, frost, sudden cold snaps (especially in spring), or attack by pests or animals.

Most of these causes either do not arise in Biodynamics or are relatively easily dealt with. Many of these potential problems simply do not occur provided we:

- Use 500 to progressively develop our soils.
- Feed plants within nature's organisation, with an adequate supply of soil humus, ensuring that the soil water remains relatively free of dissolved elements.
- Ensure that soils are well drained.
- Space plants well to allow good air circulation and good access to light.
- Prune fruit trees to allow good air circulation and sunlight.

When adverse conditions exist, such as excessive rain, low light or long cloudy periods (especially when it is warm), storm or hail damage or sudden cold snaps, we can use a range of Biodynamic techniques to restore balance and prevent disease taking hold.

Biodynamic techniques

Preparation 501

Preparation 501 is the most powerful Biodynamic tool for countering lack of light or over-lushness in plants, whether caused by extended overcast periods, excessive rain or excessively rich soil for the particular crop. Any of these conditions will predispose plants to attack by fungal diseases. The situation should be recognised and 501 applied before disease appears. See Chapter 12 for details.

Valerian

Biodynamic preparation 507, valerian, is used when sudden cold snaps occur in spring – after sap flow in fruit trees has begun – and cause a shock to the tree and interrupt the sap flow, which can lead to the development of diseases such as curly leaf in peaches and nectarines. Valerian brings warmth to the plant, overcoming the shock. It is best applied just before (ideally a few hours before) an expected sudden reversal in spring's progress, but is still helpful afterwards, for instance the morning after a frost. Valerian has proven particularly helpful for stone fruit, citrus and berries affected by cold snaps and has also proven to be a great help in recovery from storm damage.

Enough liquid valerian is added to pure warm water that one can just discern the scent. This may vary between 5 and 15 ml of valerian per 3 imperial gallons (13.5 litres) of water. The water is stirred as for 500 or 501, for 20 minutes, and then lightly sprayed as a mist over the leaves.

Horsetail/casuarina tea

Equisetum arvense (horsetail) and casuarina (she-oak) are both used as mild anti-fungal agents against mildews, rusts and similar plant diseases. Both equisetum and casuarina have a very high silica content. Equisetum and casuarina are best alternated with 501, which balances the over-lush condition that invites fungal problems.

In Australia, horsetail is a noxious weed and is not allowed to be grown. Once established, it is almost impossible to eradicate by any means. Instructions follow for readers in countries where it grows wild. For those in Australia, dried equisetum can be used but is prohibitively expensive for all but garden-scale use. Casuarina has proven to be nearly as effective as equisetum. All casuarina species may be used. They are widespread native plants in Australia and are also native to South-East Asia, India, East Africa and the islands of the western Pacific. It has also become established in the southern United States, Brazil and West Africa. It can be planted for harvesting in many other countries, though it is considered invasive in some countries.

One of the most experienced Australian users of casuarina is Tom Meredith, a Far North Queensland Biodynamic grower of fruits and vegetables. He finds that all casuarina species work well, those growing close to the sea being particularly effective owing to salt spray on the leaves. In the very challenging wet, humid tropical climate of Far North Queensland (8m annual rainfall), he finds that 501 and casuarina enable the successful growing of mildew-susceptible crops such as courgettes.

In addition to killing fungal and rust

Figure 20.3. Casuarina, a very light-influenced tree. The needle-like leaves can be seen in more detail at top right.

diseases, casuarina, if allowed to ferment for at least 3 days, also kills insects such as aphids, mites and scale, helping to reduce the insect-borne spread of many fungal, bacterial and viral diseases. Fermented casuarina has similar properties to soap, killing small insects by suffocation, but doesn't kill the beneficial predator insects, whose breathing holes are larger and aren't covered by the 'soap'.

Both equisetum and casuarina can be used on a wide range of vegetables, fruits and vines, particularly in wet or humid periods, and especially close to full moon and perigee, when there is an increased risk of fungal diseases. On vines, spray at budburst and again as the leaves begin to fall, with in-between applications if necessary, in tandem with 501, the most effective preventative of fungal diseases in a wet spring or summer. See p. 178 for *Phytophthora*.

Equisetum

Fresh – gather 1kg of fresh sterile mature summer stems (which grow after the spore-bearing stems have wilted) of *Equisetum arvense*. The sterile stems have many side shoots. Cover with pure water and bring to simmering point. Simmer gently for 40 minutes. Allow to cool, then filter and either store in a cool dark place for up to 2 weeks or use straight away. For use, dilute 1 part to 9 parts water and stir as for 500 and 501, for 20 minutes, before spraying.

Dried – use 100g of dried horsetail per 4 litres of pure water, or 25g per litre of water if only a small amount is required. Bring to simmering point and simmer lightly for 1 hour. Allow to cool and strain before use. Stir as for 500/501 for 20 minutes and spray, covering both sides of the leaves.

Undiluted equisetum decoction can be stored in a cool dark place (strained or unstrained) for up to 2 weeks before use.

Casuarina

The fresh leaves of the male casuarina plant are used. Male trees do not produce any fruiting cones; they flower annually, giving the leaves of the male trees a rusty-reddish appearance. Choose a mature tree, checking to make sure that no fruiting cones are present on or under the tree.

Three-quarters fill a copper, stainless steel or enamelled steel vessel with fresh casuarina leaves (not packed in tightly) and just cover with pure water. Heat to simmering point and simmer for several hours, until the froth disappears (this can take up to 6 hours). Allow to cool. If only using the casuarina for its anti-fungal properties, filter

and dilute to an average tea colour (typically 1 part of casuarina liquid to 10 parts of water). Stir for 20 minutes as for 500/501 before spraying. Spray both sides of the leaves.[5]

If using the casuarina for both its anti-fungal *and* insecticidal properties, allow the unfiltered decoction to ferment for at least 3 days. When it is needed, filter, dilute, stir and spray as above. The liquid can be stored, filtered or unfiltered, in a cool dark place (it can be in a plastic container) for up to 2 weeks before use.

In hot climates, Meredith advises that casuarina and equisetum should be sprayed only after 3pm to lessen the chance of the sun burning young growth and also to slow the drying process, thus making the 'soap' more effective at covering the breathing holes of pests and killing them.

In cooler climates, Frances Porter (a long-time BDAAA Biodynamic preparation-maker) advises that morning spraying is best, especially in spring. 'This also allows the Casuarina to dry on the plant foliage before nightfall.'[6]

Biodynamic winter tree spray

A nourishing and protective spray for dormant trees and bushes, comprising kaolin clay, liquid cow manure, sodium silicate and microtomic sulphur. See pp. 175–176 for details.

Situations where more than one spray is beneficial

Porter suggests that in hot tropical climates, both casuarina and 501 be used when the weather has been cloudy, humid or drizzly for several weeks. In temperate climate springs, when cold conditions bring on curly leaf, black spot or brown rot, she recommends that both casuarina and valerian (for anti-fungal activity and warmth activity, respectively) be sprayed.[7] They can be combined.

Sodium silicate

This is used as a component in Biodynamic fruit tree spray (see Chapter 23) and also by itself for diseases such as leaf curl in peaches and nectarines and shot hole in apricots. Also effective against other fungal diseases such as scab and powdery mildew. Use at 0.75% (75g per 10 litres of water) or 0.5% (50g per 10 litres of water) when leaves are still very small. Sodium silicate is a severe eye irritant: wear protective goggles and take care when handling and spraying it.

Other natural plant sprays

Yarrow

A mild anti-fungal agent that can complement and reduce the amount of sulphur sprayed in a vineyard. It strengthens grain crops. Maria Thun suggested that it has a good effect on seed vitality in the grains; spray early and also later, before harvest. Normally the flowers are used, but tea made from the fresh leaves has been found useful against leaf-spotting diseases in tomatoes.

- Use 12g dried flower heads to 4 litres of water.
- Bring to boiling point, then remove from the heat and leave to infuse for 20 minutes.
- Dilute to 40 litres – enough for 1 hectare.

Dandelion

This builds resistance to fungal diseases. Can be used alone or mixed with copper or sulphur. As with yarrow:

- Use 12g of dried flowers in 4 litres of water.
- Bring to boiling point, remove from heat and allow infusion for 20 minutes.
- Dilute to 40 litres and spray.
- This amount will cover 1 hectare.

Chamomile

This has a similar effect to yarrow and dandelion.

- Use 12g of dried flowers in 4 litres of water.
- Bring to boiling point, remove from heat and allow infusion for 20 minutes.
- Dilute to 40 litres and spray (1 hectare).

It can be mixed with copper or sulphur when used on vines and can reduce the amount of copper required.

Stinging nettle

A mild anti-fungal agent used in vineyards and on potatoes, stinging nettle is best used together with other plant teas or with copper or sulphur to strengthen its effect. It enables the amount of copper or sulphur used to be reduced. It also helps against spider mites in vineyards.

The following recipe makes enough to spray 1 hectare:

- Use 1kg of fresh nettles or 100g of dried nettles (best picked at the leaf stage or at early flowering) in 4 litres of water.
- When it comes to the boil, remove it from the heat and allow it to infuse for 20 minutes.
- Dilute to 40–50 litres and spray over 1 hectare.

A small amount of powdered clay (e.g. kaolin, at 1% by weight or 500g per 50 litres of water) can be added, but not if you are mixing the infusion with sulphur or copper.

Milk
Skimmed or whole milk is an effective, mild anti-fungal agent, especially for powdery mildew. Dilute 1:10 with water and spray plants to saturation.

Breeding vitality in seed
As mentioned above, Biodynamic seed produces stronger plants with each generation and the plants become progressively more resistant to pests and diseases.

Diseases of fruit trees
Diseases of fruit trees are covered in more detail in Chapter 23.

Allowable materials for disease control
Organic and Biodynamic growers are allowed to use a range of safe materials to counter diseases when other methods fail. The (Australian) National Standard for Organic and Bio-Dynamic

Produce list of allowable materials for plant disease control appears in Figure 20.4 (overleaf). In other countries, consult your own national organic standards.

FOOTNOTES FOR CHAPTER 20
1. Podolinsky, *Bio-Dynamic Agriculture Introductory Lectures, Vol.1*, pp. 37–38.
2. For instance, Philbrick and Gregg, *Companion Plants*.
3. See Rod Turner, 'The Inheritance of Vitality in Biodynamic Plants', *Biodynamic Growing*, 4 (2005), pp. 18–19.
4. It can be downloaded for free from https://www.agriculture.gov.au/sites/default/files/documents/national-standard-edition.pdf
5. Adding 0.5% sodium silicate can improve sticking ability.
6. Tom Meredith, 'Casuarina Tea for Fungal and Insect Problems', *Biodynamic Growing*, 19 (2012), p. 30.
7. BDAAA, *Biodynamics A Practical Introduction*, Powelltown, Victoria, 2011.

National Standard for Organic and Bio-Dynamic Produce – Edition 3.8

Table A3 Plant disease control

Substances	Specific conditions/restrictions
Ayurvedic preparations	None
Biological controls	Naturally occurring cultured organisms only
Copper e.g. Bordeaux and Burgundy mixture	Annual copper application must be less than 6 Kg/Ha/annum on an average rolling basis
Essential oils, plant oils and extracts	None
Granulose virus preparations	Need recognised by certification organisation.
Homeopathic preparations	None
Light mineral oils (such as paraffin)	None
Lime	None
Lime-sulphur	None
Natural plant extracts excluding tobacco	Obtained by infusion and/or made by the farmer without additional concentration
Potassium permanganate	None
Potassium soap (soft soap)	None
Propolis	None
Seaweed, seaweed meal, seaweed extracts	None
Sea salts and salty water	None
Skim milk or skim milk powder	None
Sodium bicarbonate	None
Sodium silicate (water-glass)	None
Sulphur	In wettable or dry form only
Vegetable oils	None
Vinegar	None

Figure 20.4. Appendix C, Table A3.
National Standard for Organic and Bio-Dynamic Produce, Edition 3.8: November 2022, Department of Agriculture, Fisheries and Forestry, Canberra. Reproduced here under a Creative Commons Attribution 4.0 International Licence. No changes have been made. Copyright is owned by the Commonwealth of Australia. Disclaimer: The Australian Government acting through the Department of Agriculture, Fisheries and Forestry has exercised due care and skill in preparing and compiling the information and data in this publication. Notwithstanding, the Department of Agriculture, Fisheries and Forestry, its employees and advisers disclaim all liability, including liability for negligence and for any loss, damage, injury, expense or cost incurred by any person as a result of accessing, using or relying on any of the information or data in this publication to the maximum extent permitted by law.

Chapter 21

Weed Management

Nature is always working towards harmony and balance. We are really only at the very early stages of comprehending the complex and highly organised interrelationships in nature. This striving towards harmony and balance is very obvious with regard to soil and plants.

When soil is left bare, nature brings various plants to cover the soil and protect and nurture it. When soils are unbalanced, certain weeds come, either as a result of the imbalance or to redress the imbalance. Acid soils often see sorrel growing; docks thrive in poorly drained soils and soils with excess nitrogen. Compaction brings weeds with deep taproots such as thistles and flatweed that strive to penetrate and open up the compaction. Deep-rooting weeds also bring minerals up from deep in the subsoil, remedying various deficiencies.

Where there are mineral deficiencies, specific weeds grow that concentrate deficient minerals either by bringing these minerals up from deep in the subsoil or via biological transmutation (see pp. 25–27). For instance, in many parts of Australia, if copper levels in the soil become too low, heliotrope ('potato weed') grows. Heliotrope accumulates copper and,

if handled correctly by the farmer (ploughed in as it starts to flower or at least before it dries off), can restore the copper levels in the soil. Capeweed, apart from its role in quickly covering bare soil, accumulates calcium and thereby can help remedy a deficiency.

When plants grow where we don't want them, we call them 'weeds'. While recognising the roles that weeds can play in nature, we have to learn to intelligently manage them so they don't adversely affect the crops we want to grow.

General Weed Management Strategies

There are many different techniques we can use.

Weeds as green manure

This technique recognises the fact that many weeds come specifically to rectify imbalances in the soil. This process becomes particularly highly developed under the influence of the Biodynamic preparations: 'This happens via the "sense" our Biodynamic preparations activate.'[1] The weeds that come are allowed to grow and worked into the soil as they start to flower.[2] Not only does this allow many of the weed seeds in the soil to germinate, thus reducing subsequent weed germination, but it ensures that many mineral deficiencies are rectified by the weeds that nature chooses to grow. This process was objectively observed, and its practical application developed and proven, in extensive field application by Biodynamic farmer Barry Edwards in western Victoria.[3]

Edwards fallows with a one-way disc in the late winter or early spring before sowing a crop, ploughing in as much bulk as possible (grasses, clovers, weeds, everything). The area is worked again with tines in early summer. Summer rains germinate more weeds, which are allowed to grow until they start to flower or are just beginning to dry off, whereupon they are ploughed in. Although the growth over summer reduces soil moisture somewhat (as opposed to a bare fallow), they bring a large gain in nutrients and organic matter, which is converted into colloidal humus by the soil biology, resulting in the autumn-sown crop being able to utilise growing season moisture much more efficiently. The gain in nutrients and humus leads to cleaner crops because the weeds don't need to grow to redress imbalances.

The exact mechanism by which weeds accumulate particular elements in soils that are severely deficient in those elements is not yet fully understood. Some elements are retrieved from deep in the subsoil and some are absorbed from the air. Moreover, there is growing evidence of 'transmutation of elements' (see pp. 25–27).[4] Once the weeds have been allowed to fully express themselves and been worked in, they generally do not need to grow there again for some time. Using weeds as green manure also works well with garden beds that have become weed infested by neglect. Don't pull the weeds out (unless you need to plant a new crop immediately) and thus remove all the minerals and organic matter nature has developed; instead chop them up and dig them into the soil where they belong. Ensure that enough time is allowed for decomposition before you plant the next crop.

False seed beds

Soil is firstly prepared for a crop: this may include primary and secondary cultivation, scarifying, chopping up and working in green manures, working in compost, and bed forming (if crops are to be grown in raised beds). Having prepared the soil, we allow time for weeds to germinate and then we cultivate to destroy them just before or just after they emerge from the soil. If the weather is dry, irrigate if possible, to encourage weed germination. If you have enough time, and especially if you think there is a lot of weed seed in the soil, allow a second crop of weeds to germinate, and then cultivate again. This can even be done a third time before planting if there is time. Creating and cultivating false seed beds will considerably reduce the number of weeds that will germinate after the crop is sown or planted. It is much easier to cultivate weeds before the crop is planted than after.

The same principle applies whether we are growing vegetables, grains or other crops, though with grain crops weed reduction techniques begin a year or even two before the grain crop and there is often little time for false seed beds, since sowing must be timed to soil moisture levels and the seasonal progression towards winter.

Pre-emergent flame (or steam) weeding

Prepare your seed bed a week or two before sowing the crop. Weeds will germinate and appear before the crop. If the conditions are dry, and irrigation is available, applying some water will quickly bring a flush of weeds. A quick pass with a hand or tractor-mounted flame weeder will 'cook' the delicate, newly emerging weeds before the crop emerges. This can be done before sowing the crop, but can also be done after sowing, since the weeds tend to emerge ahead of the crop. The procedure obviously requires careful timing to avoid damaging the about-to-emerge crop.

Weed control after planting or germination

After the crop has germinated or been transplanted, weed control must be carried out vigilantly and in a timely manner, killing the weeds as early as possible after germination. Soil can regularly be lightly 'tickled' (lightly cultivated). This not only kills many weeds at or before their emergence, but maintains a 'soil mulch' that conserves soil moisture.

Various implements are available. Some destroy the weeds by loosening the soil and disrupting delicate root systems. These implements are most effective at or before weed emergence. Once weeds are more established (but still small), cultivation works best when conditions are drying and some sun warmth is present to dry the weeds out. Cultivating in cold, damp conditions is not very effective with established weeds, since they can easily regrow.

Some implements cut the weeds under the soil. These implements are effective at any time, including when cool, damp conditions prevail.

Flame weeders can be used to kill inter-row weeds, provided they are fitted with shields that protect the crop.

Some implements kill weeds in the inter-row (either by cutting or tickling) and at the same time throw a little soil into the crop row, smothering weeds there. Fix Engineering's Weed-Fix powered weeder does this, as do various types of blade or wheel incorporated in other implements.

A Biodynamic technique – 'peppering'

In his 1924 lecture series, Rudolf Steiner suggested a method of controlling weeds over a period of time. Mature weed seeds are collected and burnt on a very hot fire. The resulting ash is scattered over the area to be treated. The ash can be distributed on its own or mixed with seed, lime, rock dust or any other soil amendments that are being spread.

Steiner suggested that if the weed seed ash were spread annually for 4 years, the weeds would become progressively weaker and weaker. He didn't specify a particular phase of the moon for the seed burning. Various Biodynamic practitioners have tried full moon or new moon, but no conclusive evidence for any particular time has yet emerged. A commonly used practice is to collect and burn the weed seeds at or just before full moon and spread the ashes at or just before new moon. Maria Thun recommends burning different weed species when the moon is in a zodiac constellation particular to each species. This is summarised in Pierre Masson's book *A Biodynamic Manual*.[5] The list is not, however, extensive. More research is needed.

One of the most successful recorded uses of the peppering technique has been by New South Wales Biodynamic farmer Kevin Terlich, who burnt Bathurst burr seed to ash in an oil-fired furnace (with air blown through a jet nozzle) used for making steel points for rippers. The seeds were put in a vessel designed to hold molten steel and burnt at a temperature of around 900ºC. The resulting ash was mixed with the wheat seed before sowing. The result was an enormous reduction in Bathurst burr in the crop. Kevin's son, Trevor, thinks his father (now deceased) was experimenting with collecting and burning the seed at full moon or new moon but is unsure of the details. However, Kevin, Trevor and Alex Podolinsky all agreed that the very high temperature of the furnace was probably an important factor in the immediate and dramatic results. See Chapter 22 for more detail.

Weeds in Pasture

In pasture, most conventional farmers want only a limited number of species, rye grass being the favourite (and often the only species wanted by conventional farmers) in dairying areas in southern Victoria. Clover is feared because water-soluble-fed clover can kill cattle in spring via bloat. (I woke one morning on a conventional dairy farm where I milked, to find 14 cows dead from bloat in the paddock next to the house.) Regular nitrogen applications force rapid voluminous growth in the rye grass and unnaturally large volumes of (unhealthy) milk. Biodynamic farmers also like rye grass, but they also want abundant clover to supply natural nitrogen – and it is very rare to see a case

of bloat on a Biodynamic farm. They also want many other species in the pasture, as many as 40 species being regarded as a good mixed pasture. Some of these would normally be regarded as weeds by conventional farmers. They would not be eaten by stock on a conventional farm, indeed some may cause health problems, but they are relished and provide good varied nutrition and health benefits on a Biodynamic farm. Animals have highly developed instincts and will seek out particular species to provide nutrients they need or to remedy a health problem. Cows will even seek particular species to eat to remedy a health problem in their calf via their milk.

Under Biodynamic management, not only do many classes of weeds – that came to balance deficiencies or to open up tight soils – progressively disappear, but weeds in general become progressively easier to deal with. As the soil becomes more friable and crumbly, weeds become easier to cultivate or pull out by hand. They also become much more palatable to stock and in some cases less dangerous. For instance, rye grass is a serious weed in grain crops. On the Western Australian broadacre property of John and Bernadette Cashmore, who have now retired, the sheep ate the rye grass seeds and it was not a problem in grain crops. This does not happen on conventional properties. Similarly, wild radish is a serious grain crop weed because the seeds are the same size as the grain and can't be separated after harvest. The Cashmores had very few wild radishes in their pasture, and those that did venture up were relished by the sheep. By contrast, their conventional neighbour had a big problem: after herbicides had been sprayed on a field full of wild radish in preparation for a crop, the radish came back even harder, and the sheep wouldn't touch it, since it was unpalatable. Capeweed on conventional farms can cause nitrate poisoning in pigs, cattle, sheep and horses but is relished by stock on Biodynamic farms and causes no problems there.

Capeweed comes and goes seasonally. A very dry summer/autumn can bring it in abundance. One day while topping a spring field with plenty of capeweed, I was stopped by an elderly neighbour who said I must spray the capeweed because the whole field would become infested and have to be re-sown. He, of course, sprayed annually, only to see the capeweed return every year. I explained that we were certified Biodynamic and didn't use chemical sprays. He went away shaking his head. Next spring, we had virtually no capeweed, just beautiful rye grass, clover and many other useful species. I noticed the neighbour looking over the fence thoughtfully one spring day.

On our South Gippsland (Victoria) property, thistles became a rarity after a few years of Biodynamic applications. They were primarily present to ease compaction with their deep taproot. In our naturally rich Dumbalk Valley soils, mineral deficiencies were not really an issue; in our case, compaction was the main reason for the presence of thistles. As the soil developed a more open, friable structure with prepared 500 use, the thistles disappeared. In some very heavy clay soils, thistles are valued and used as green manure by Biodynamic farmers and continue to serve a useful purpose.

Crop rotation has an important influence on weed problems. If we try to grow the same crop over and over again on the same spot, deficiencies or imbalances will inevitably develop over time, and weeds particularly suited to the deficiency will tend to become more and more of a problem. This does not apply to permanent pasture, which is the most healthy and balanced community of plants that we can grow.

Re-sowing a pasture will not eliminate undesirable species if the underlying conditions or deficiencies have not been addressed. For instance, in southern Victorian pastures, sweet vernal thrives on potassium- and phosphorus-deficient soils and is also encouraged by poor soil structure and waterlogging. It is not a very nutritious or desirable species, though it is welcome as a small component of a widely varied orchestra of plants and one that gives a wonderful aroma to the pasture. If the underlying conditions are not addressed, it will continue to flourish after re-sowing.

Weed Control in Grain Crops

The following points are from Dayle and Terri Lloyd at Dumbleyung, Western Australia:

Seed reduction in pasture breaks

Biodynamic grain growers generally use a crop rotation system that includes a pasture break of 2–4 years between grain crops. For instance, Dayle allows a 2–3-year pasture break between grain crops. In the first year after the crop, the grass is allowed to fully develop and set seed, but for a year or two afterwards the grass is grazed more intensively and also periodically slashed. This gives good control of grass seeds before the crop is grown. Cutting hay the year before sowing a crop also helps.

Cultivation and secondary cultivation in a range of soil types

Buster points on scarifiers can get the chipping effect of a chisel plough and achieve good turning of the soil. Offset discs also work well but are expensive. Tractor speed is adjusted according to soil type and moisture levels: a wetter soil with more grass on top needs a higher speed to get inversion of the sod (6–7 kmph). Drier sod with less green material needs a slower speed (5.5–6.5 kmph). Heavy or rotary harrows can help free large cloddy weeds or wild oats that may germinate after seeding but before crop emergence.

Secondary cultivation usually needs a faster speed (7–8 kmph) to get good weed control and prepare a finer seed bed. Leaving moist soil soft, dry and fluffy on top after seeding discourages weed germination, but soils that are too dry may be vulnerable to wind erosion. Always remain vigilant for heavy germination of weeds and be prepared to stop seeding and re-cultivate or harrow if needed. In dry conditions you can use the dew of the early morning to work the soil.

Implements

Gardens and small market gardens

Three-pronged hoe

A good general purpose cultivating implement that 'tickles' the soil, destroying newly germinated weeds. It is, however, more effective if the middle prong is removed, since the hoe

tends to drag soil too much otherwise. I use two of these implements, one with two prongs, and one with only the central prong, which is good for cultivating between closely planted vegetables in the row.

Figure 21.1. Small flame weeder.

Figure 21.2. Gung hoe.
Photo: Allsun Farm (www.allsun.com.au).

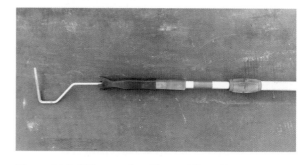

Figure 21.3. Wire weeder. Photo: Allsun Farm.

Figure 21.4. Wheel hoe. Many attachments are available.

Figure 21.5. Trapezoidal hoe. Photo: Allsun Farm.

Figure 21.6. Oscillating stirrup hoe.

Figure 21.7. Flame weeding, Taucha, Germany. Photo: Maria Bienert.

Gung hoe

A sharp cutting draw hoe designed by Eliot Coleman, author of *The New Organic Grower.*

Oscillating stirrup hoe

A sharp cutting hoe that is used with a forward-and-back movement, cutting weeds on both forward and backward strokes.

French wire weeder

A light, sharp hoe, suitable for cutting very small weeds, that can be used in the row as well as between rows.

Trapezoidal hoe

A heavier hoe that is suitable for chopping larger weeds that have been left too long.

Wheel hoes

Wheel hoes greatly increase the amount of weeding that can be done by one person and make it easier and much more efficient. They are very useful in small market gardens.

Flame weeders

Small units can be obtained that are especially useful in small market gardens.

Farm implements

Simple cultivators

These include:

- Scarifiers for cultivating the whole area before sowing or planting.
- Comb harrows, which have very thin tines that are drawn through the soil before or

Figure 21.8. Row crop weeder with cutting blades, Agrilatina, Italy.

Figure 21.9. Detail from Figure 21.8 showing cutting blades and crop shields.

after the emergence of cereal crops. On some comb harrows (e.g. the Hatzenbichler) the degree of 'aggression' in the tines can be set so as to dislodge the weeds but not the crop.

- Triple-K-tined cultivators or 'scufflers', S-tines or curly tines, which can be used over the whole area before sowing or planting or can be set up to cultivate only between the crop rows. Row guards can also be fitted to protect the crop.

Cutting implements

A wide range of tractor-mounted attachments are available for cutting weeds just under the soil surface.

Powered implements

These include the Weed-Fix implement manufactured by Michael Fix,[6] and various powered brushes.

Tractor-mounted flame weeders

These generally using liquid propane gas as the fuel. They can be used before the crop is planted,

Figure 21.10. Comb harrows, market garden. Photo: Agrilatina Cooperative, Italy.

after sowing but before the crop emerges, or in a growing crop if crop shields are fitted to protect the growing plants.

Control of Some Specific Weeds

There are so many varieties of weeds in different areas and different countries. The following is just a sample list with some control strategies.

Figure 21.11. Weed Fix – powered rotating fingers, crop shields. Photo: Fix Engineering (www.fixengineering.com.au).

Figure 21.12. Triple K cultivating weeds before planting, Harm family farm, Queensland. Photo: Harm family.

Figure 21.13. Triple K cultivating weeds after planting brassicas, Harm family farm, Queensland. Photo: Harm family.

Local research and trial and error will find solutions to most local weed problems. Some growers have tried peppering using the vegetative parts of, for example, couch, but, so far, results have been inconclusive.

Bent grass (*Agrostis spp.*)

This is invasive in acid (and often low-nitrogen) soils in high-rainfall southern Victorian areas. It is considered undesirable in pastures because it inhibits the growth of clover. Podolinsky qualifies its bad reputation by noting that it is 'good tucker for calves'. Some success has been achieved by burying it deeply with a mouldboard plough at the beginning of winter. Otherwise, cultivating repeatedly during hot, dry periods will dry out and kill bent grass. If soil conditions conducive to its growth are not remedied, it will return, germinating from seed left in the soil.

Cape tulip (*Moraea miniata* and *Moraea flaccida*)

This is a toxic weed, particularly for cattle. It can be controlled by, first, slashing before it sets seed, and then cultivating to expose the bulb to be dried out by the sun. In some cases bulls help, since they pull it out, bulb and all, in winter when the soil is moist, and then spit it out.

Couch grass (*Cynodon dactylon*)

This can be greatly reduced by repeated cultivation during hot, dry periods. In gardens, dig it out with a garden fork, shake all earth off the roots and leave the plant in the sun to dry out. Couch is one of the most persistent problem weeds. Any tiny piece of couch left in the soil, even as deep as 90cm, can survive and regrow for up to 2 years.

Kikuyu (*Pennisetum clandestinum*)

This is quite a useful pasture grass in some areas, but can be a real nuisance in gardens and orchards. It can be eradicated from pasture while preparing ground for vegetables or fruit trees, by repeatedly cultivating during hot, dry weather. In South Australia and southern Victoria, kikuyu doesn't produce seeds and so is easier to eradicate. In gardens, cut the soil into workable-size portions with a sharp spade, dig up with a garden fork and bash the soil off the roots. Leave to dry out in the sun.

Blackberries

Goats are a very good way to control blackberries, since they love eating the leaves. Goats also help control gorse and eat the flower heads of ragwort, flatweed, capeweed and thistles. Goats need to be managed carefully to avoid damage to fruit trees, crops and gardens.

Ragwort

This is toxic to cattle but not to sheep. Cut the flower heads off ragwort that is flowering, even if the flower buds have not yet opened (put the flower heads in a plastic bag and dispose of them carefully) and dig the whole plant up with a fork, taking care not to break any roots if possible; roots that break off can regrow. Friable soil structure facilitates removal of the plant without breaking roots. Hang the plant on a tree branch or fence line to die. In badly infested country, sheep are a better option than cattle. Sheep will eat the flowers too.

Thistles

Often a response to poor fertility and/or soil compaction, thistles tend to disappear as soil structure and fertility improve with annual (or twice annual if possible) prepared 500 applications. In the meantime, slashing before flowering can help reduce seeding and some stock can eat the cut, wilted thistles before they dry out. Hand hoeing (just under the soil surface) will suffice if only small numbers are appearing.

Some farmers argue that it is best to allow thistles to seed before slashing, since then the maximum possible mass of roots and tops will have developed (as a green manure). This may seem risky, but in fact, if the thistles have been allowed to do their job fully, there may be no need for them to grow in future, despite there being abundant seed.

On very heavy clay country, thistles can be a recurring presence and their deep taproot plays a very useful role in keeping the heavy soil open. Some Biodynamic farmers on these heavy clays welcome them as a valuable green manure species.

FOOTNOTES FOR CHAPTER 21
1.   Alex Podolinsky, *Biodynamic Growing*, 9 (2007), p. 32.
2.   Barry Edwards, in western Victoria, finds that even if the weeds produce seed, provided they are worked into the soil while they are still green, the balancing effect still works and the seed does not cause problems.
3.   Barry Edwards, 'Weeds, Just What Your Soil Needs', *Biodynamic Growing*, 9, 2007, pp. 31–33.
4.   See pp. 25–27 and also Frances Porter, 'Old Studies Have Much to Teach Us', *Biodynamic Growing*, 2 (2004), pp. 11–13.
5.   Pierre Masson, *A Biodynamic Manual*, Floris Books, Edinburgh, 2011.
6.   www.fixengineering.com.au

Chapter 22

Peppering

In his 1924 lecture series, Rudolf Steiner described a unique method of countering weeds, insects and animal pests. This involved collecting the weed seeds, whole insects or animal skins, burning them and distributing the ash over the area to be treated. He said that this process would reduce the prevalence of the weeds and insects over a period of 4 years and immediately repel the 'ashed' animal pest. Experience in Australia suggests that insects and weeds can also be deterred immediately after application of the ash or charcoal, provided certain conditions are met. Steiner described the spreading of the ash as akin to 'sprinkling pepper', and the process is now widely referred to as 'peppering'.

Steiner suggested specific times when this process could be carried out. In the case of animals, the burning of the skin and scattering of the ash were to be done when Venus was in the constellation of Scorpio (i.e. when Scorpio is behind Venus in the sky; this varies from year to year, but occurs over a roughly 3-week period between October and January). For insects he suggested that the best time would be while the sun is moving from Aquarius towards Cancer as the insect passes from the winged state towards the larval state; he specified

Taurus as the most effective time to pepper the winged insect. He advised that testing was required to refine his suggestions. Steiner didn't specify any particular times for the ashing and peppering of weeds.

Peppering is one area of Biodynamics that is still in its early stages. More research is needed to verify Steiner's suggestions and to refine the method of application. In the northern hemisphere, German Biodynamic researcher Maria Thun carried out many trials and made specific recommendations. However, there is, at this stage, a lack of corroborating studies that would give us sufficient confidence to fully endorse her recommendations. Furthermore, her recommendations were based on northern hemisphere research.[1] For insects, Steiner's recommended times begin in the southern hemisphere temperate climate autumn and continue through the winter; this is not very helpful for countering insects in the more prolific growing times of spring and summer.

In the southern hemisphere, some findings are emerging. Many Australian Biodynamic farmers have reported success in applying peppering against rabbits (a major farm pest in Australia). Rabbit skins are collected, dried (stretched over a length of number 8 wire bent into a U-shape and hung up under cover to dry) and stored in sealed plastic bags until Venus is in Scorpio, whereupon they are burnt in a hot wood fire and the ashes or charcoal peppered. The pepper can be spread over a general area or applied around the boundary of an area from which rabbits are to be excluded.

As mentioned in Chapter 21, New South Wales Biodynamic farmer Kevin Terlich reported a massive reduction in Bathurst burr in a wheat crop after peppering. He burnt the weed seed in an oil-fired furnace at 900°C, an extremely high temperature, and mixed the resulting ash with the wheat seed before sowing.

North Queensland farmer Tom Meredith has applied peppering extensively, carried out controlled experiments and developed some specific refinements.[2] He stresses the need to burn the weed seeds, insects or animal skins at a high temperature, without air, to produce a shiny charcoal rather than ash. Ash is produced when the burning is done *with* air.

Meredith carried out a controlled experiment using peppers to counter grasshoppers in a particularly bad year, when the ground was covered with the insects, which completely devoured any transplanted seedlings. Grasshoppers were collected, frozen and divided into 10-g lots for the experiment. He compared amaranth seedlings peppered with the sun in Aries (the 'wrong' time) with amaranth seedlings peppered with the sun in Taurus (the 'correct' time). The other variable tested was the length of time the insects were left in the hot wood fire. In each case, one control tray of seedlings was not peppered.

The results were conclusive. The control trays (not peppered) were completely devoured, while a steadily increasing degree of protection was afforded to the seedling trays that had peppers applied that were produced by grasshoppers being left in a cast iron camp oven in the hot wood fire for 1, 2 or 3 hours, or all night; the material left in the fire all night afforded the greatest degree of protection. The trays peppered with the sun in Taurus were consistently better protected in each instance than were the trays peppered with the sun in Aries. However, the seedlings peppered with the sun in Aries, and the camp oven left in the fire all night, were sufficiently protected that a commercial crop was obtainable. This is highly significant because, in Meredith's tropical climate, the sun is in Taurus at a prime growing time. In cooler southern climates, sun in Taurus (late autumn or early winter) is not a main cropping time except for grains and some winter crops. Alternative times must be tried, to protect crops over the warmer spring, summer and early autumn growing times.

Based on Meredith's experience and experimentation, together with other sources, I recommend the following:

- Collect the weed seeds, insects or animals at any time, dry or freeze them (freezing is best for insects and small animals) and store them until the correct astrological time for burning.
- Weed seeds should be burnt between new moon and full moon. Although some people recommend full moon, others recommend new moon; more research is needed. Where

weeds produce tubers or bulbs (e.g. nutgrass) or coppice (regrow from the base) or send roots down from nodes as in running grasses, Meredith burns these parts along with the seeds. He tries to burn them when the moon is in the appropriate zodiac sign if applicable. For example, nutgrass is burnt at a root time (Virgo, Capricorn or Taurus), though whether this makes much difference remains uncertain.

• For fully developed insects such as grasshoppers, the best time for charcoaling and peppering is when the sun is in Taurus (currently mid-May until mid-June). This is supported by Thun's research. However, as indicated above, burning at the 'wrong' time is still sufficiently effective provided the vessel is left in the hot fire overnight.

• For peppering watery pests such as insect larvae and molluscs such as slugs and snails, sun in Cancer (mid-July to mid-August) is generally recommended. Again, however, in cooler southern hemisphere climates slugs and snails are less active at this time, and a later time can be tried – for instance, when the sun is in Scorpio (the succeeding water sign to Cancer), from late spring to early summer. Rod Turner, a Melbourne Biodynamic gardener tried this, burning 30 adult slugs and ten snails. As the amount of ash obtained was so small, he trialled adding the ash to water and stirring as for 500 before spraying the liquid over the garden. The results were not conclusive, but numbers were probably reduced. More research is needed.

• Most Biodynamic farmers burn the whole skin of animal pests such as rabbits. Meredith's recommendation is that small animals such as mice, rats and small birds should be burned whole, while for larger animals the internal organs (except the stomach and intestines, which are discarded) should be burned together with a strip of fur from along the backbone to the tail. Meredith stores the animals in a freezer until required.

• The hotter the fire, the better. This ensures a shiny charcoal, the most effective for peppering in Meredith's experience. Make a large wood fire, put the substance to be peppered in a cast iron camp oven (best) or metal tin with metal lid, ensuring that there is a vent in the lid to prevent the vessel exploding. Leave this in the fire until it has burnt out. Grind the charcoal as finely as possible before spreading.

• Coat seeds at sowing time with a mix of all the peppers of pests or weeds that tend to affect them, or spread the peppers over the areas to be protected.

Although peppering usually employs the ashes or charcoal in a dry form – which may be mixed with bulking materials such as sand or basalt dust – some Biodynamic practitioners prepare homeopathic dilutions (usually to D8) and spray them over the area to be treated. As this method is little used in Australia, I refer the interested reader to Pierre Masson's *A Biodynamic Manual* for further information.[3]

FOOTNOTES FOR CHAPTER 22
1. Those interested in her specific recommendations can find a summary, together with other European recommendations, in Masson, *A Biodynamic Manual*, pp 177–188.
2. See Tom Meredith, 'Charcoaling Animal, Insect and Weed Pests', *Biodynamic Growing*, 18 (2012), pp. 10–12.
3. Masson, *A Biodynamic Manual*, Floris Books.

Chapter 23

Biodynamic Fruit Tree Care

The fundamentals of Biodynamic fruit production have much in common with those of conventional fruit production – site selection, drainage, spacing, pruning and so on. The most important difference lies in the mode of providing nutrition to the plants. Biodynamic growers use various means (including the all-important 500 applications) to build soil humus levels and open, friable soil structure. Plants feed from this soil humus as directed by sun warmth and light. Plants feeding naturally in this way are metabolically balanced and less susceptible to pests and diseases. If any remedial action is required, the Biodynamic light spray, 501, and other natural materials are used. The fruit produced is of superior flavour and keeping quality. Conventional growers supply water-soluble nutrients via the soil water. This removes the plants from nature's organisation, resulting in forced, unnatural, unbalanced growth. Such plants are more susceptible to pests and diseases and need to be sprayed with toxic chemicals to allow a crop to be produced. The fruit has poorer flavour and shorter shelf life.

Some General Considerations

Site selection

We are often limited by our land, but, if you can, choose a site that is reasonably sheltered, in full sun for most of the day and, above all, well drained. If possible, site the orchard away from gum trees (eucalyptus), which are greedy competitors and may increase the risk of *Phytophthora* root rot. However, some people do grow good fruit near gum trees.

Site preparation

Ideally, the whole area should be prepared well in advance by cultivating and sowing a green manure crop. Several green manure crops in succession would be even better. Fertilise the green manure (if necessary) with organically allowable materials such as compost, manure, blood and bone, guano or RPR (though RPR takes some time to release phosphorus). The green manures should be incorporated into the soil before seeding and before they become too woody. Spray prepared 500 at sowing and when incorporating the green manure.

If the subsoil is compacted, deep ripping should be carried out when the subsoil is semi-

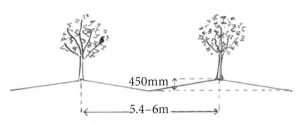

Figure 23.1.

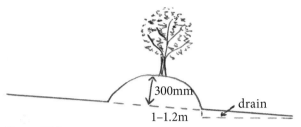

Figure 23.2.

dry, so that widespread cracking will occur. Check, by digging, to see how far the cracking extends and determine how close the rip lines should be. The Biodynamic Rehabilitator plough will relieve subsoil compaction at the same time as ploughing. On a garden scale, soil can be removed to fork depth and the subsoil loosened with the fork or with a crowbar, before the topsoil is replaced.

Before planting, measure and mark out where the rows will go. Generally, rows 5.4–6m apart will suit most fruit trees, but many factors need to be considered, including rootstock, size of trees, and pruning method. The rows should generally run downhill to provide better drainage, and the ground should be formed so that the rows are around 450mm higher (see Fig. 23.1) than the line midway between the rows. Ploughing on with discs, a Merbein plough or grading will achieve this. In rare cases where soils are deep, friable and free draining, trees can be planted on the flat without raising rows. All work with the soil should be done when it is not too wet or too dry.

If you are only planting a few trees, an alternative is to make individual mounds for the trees. Remove grass from a circle at least 1m in diameter. Loosen the soil to fork depth with a garden fork. If the subsoil is compacted, carefully remove the topsoil and work the fork or a crowbar into the subsoil and lever it back and forth to loosen the subsoil, then replace the topsoil. Incorporate some compost. Then bring in a few wheelbarrows of extra soil to build a mound at least 300mm high. A trench can be dug leading downhill away from the mound to improve drainage.

Figure 23.3. Neville Mock in his family's 50-acre (20-hectare) Biodynamic apple orchard, Red Hill, Victoria.

Figure 23.4. Biodynamic mango trees in full flower, Carnarvon, Western Australia.

If you have planted an orchard on individual mounds, you can grow vegetables in between the trees for a few years, gradually throwing soil up to form the raised rows – or get someone in to plough on with discs or similar to form the raised rows. It is best to do this in the first few years to minimise root damage.

Ideally, buy young trees. In the case of deciduous trees, whipsticks are best, though of course fruit comes earlier on older trees. Generally avoid multi-grafts (trees with two or more varieties grafted on to the one rootstock). For full vitality, a tree is best allowed to function as a unified whole, at least as far as the upper, fruiting part goes.

Decide how close together you want your trees in the row. Don't forget how important air circulation is. For stone fruit I would suggest 5.4–6m apart in the row, whereas apples and pears can go a little closer, say 4.8m, and citrus as close as 3.6m; some Biodynamic citrus growers leave as much as 7.2m between the rows. Plant deciduous trees in winter, the earlier the better. Citrus are best planted in early spring or early autumn.

Trim any broken roots off bare-rooted trees, prepare a hole of a suitable size and place the tree in the hole so that the soil surface will be at the same level as it was in the nursery. Ensure that the graft is well above the soil surface.

Fill in, making sure no air gaps remain around the roots. Firm in well, treading the soil around the tree. Water in well to settle the soil around the roots. Strong stakes (50 × 50mm) may be advisable in windy areas to support the young trees. These can be removed after a few years. On windy sites, plant each tree so that the strongest branch faces the prevailing winds.

Citrus trees are usually bought in pots or soil bags. Gently tease out any circling roots, but otherwise don't disturb the roots. Plant so that the soil surface is at the same level as it was in the pot.

Winter Spraying and Pruning

Early in the second month of winter, on a sunny day, spray cherries with dormant oil (winter oil) and other deciduous fruit trees with lime sulphur. Prune deciduous fruit trees the following week and, as soon as possible afterwards, spray with the Biodynamic winter tree spray. Ensure the spray is applied well before budburst. Cover the whole tree. Scrape or wire-brush off any loose or flaking bark before applying the spray. As the kaolin clay settles fairly quickly, the spray tank should be agitated during spraying. This spray is nourishing and healing to the bark and helps heal pruning cuts. It also protects against disease and coats the bark with a thin film that suffocates mites and kills the overwintering eggs of some insect pests. When I follow this winter routine, curly leaf on peaches and nectarines is never a problem, even in the coldest, wettest spring.

Pruning styles vary somewhat, but stone fruits do better in a vase or open-centred shape, whereas apples and pears can be vase shaped or pruned to a central leader. Citrus don't normally need much (if any) pruning.

Biodynamic Winter Tree Spray

For one unit – enough for four large or eight small trees:

- Put a heaped bucket full of fresh cow manure in a low-density hessian bag and tie the top securely. Put this in a large vessel such as a large plastic bucket or copper.
- Add 20 litres (two buckets) of water.
- Knead the bag thoroughly by hand or by poking with a blunt object until the water takes on a strong colour. Work the bag for several minutes to obtain a good strong mixture.
- Squeeze as much water as possible out of the bag as you remove it. Add more water if necessary to make up the mixture to 20 litres.
- Strain through a cloth (e.g. cheesecloth) to remove fibrous material that will block sprayer nozzles.
- Stirring vigorously with a stick or drill-driven paint stirrer, slowly add 1.5kg of kaolin clay, 400g of microtomic sulphur (wettable sulphur) and 400ml of sodium silicate.

Safety Alert
Sodium silicate is a severe eye irritant. Wear safety goggles while handling, stirring and applying the spray. Wash hands and face carefully after spraying. In the event of accident, flush eyes copiously with water and seek medical advice if necessary.

If spraying close to budburst, reduce the sodium silicate to 140ml.
- Continue stirring until the mixture is thoroughly mixed with no lumps. Strain through a cloth again; tie the cloth around the rim of a bucket, letting the cloth hang down into the bucket. Pour in a circular motion to continually disturb the sediment so the cloth doesn't block up in one spot.

- Spray over fruit trees – *wear safety goggles!* – making sure all parts are covered. Use any type of sprayer that has not had chemicals in it, but don't use your 500/501 sprayer.
- Wash all equipment well with water after use.
- The mixture will keep for a few weeks and can be reused after stirring and straining.
- For commercial orchard applications, various mechanical means have been developed to mix the materials and to keep them agitated in the tank while spraying to avoid settling. Rubber diaphragm pumps and ceramic nozzles are best, since kaolin is quite abrasive.

For large-scale application, per 100 litres water:

- Mix 20 litres of fresh cow manure mixed into 100 litres of water. For very large applications, the amount of cow manure can be reduced to as little as 20 litres per 200 litres of water. Let the mixture sit for a few days, then stir and strain through fine hessian or shade cloth to remove solids.
- Mix 5kg of kaolin clay (a powder) in enough water to make a thin soup, then add this to the liquid cow manure.
- Stirring vigorously, add 2kg of microtomic sulphur and 2kg of sodium silicate –*protect your eyes!*
- If you are spraying close to budburst, the sodium silicate should be reduced to 0.7% of the liquid (700g per 100 litres).

Tree Paste for Trunk Wounds

This is used for damaged or splitting bark, on large pruning cuts and after cutting tree galls or gummosis back to clean tissue. Cut away any loose bark before applying.

Mix equal parts of kaolin clay and fresh cow manure. Add 1% casuarina or equisetum tea that has been stirred (as for 500) for 20 minutes. To make the tea, three-quarters fill a container with fresh leaves, cover with rainwater and simmer for a few hours (or use 100g of dried equisetum leaves to 4 litres of rainwater). Sodium silicate can be substituted for casuarina/equisetum at the higher rate of 2%, but is not as effective. Apply the paste to the wound and cover with strips of hessian 'bandage'. Keep the paste and bandage moist during the healing of the wound.

Root Dip

Prepare a soupy mix of equal parts cow manure and subsoil loam, casuarina tea (1 part to 20 parts water) and some stirred 500. Roots of trees, shrubs, vines or seedlings can be dipped in this before transplanting.

The Early Years

In the early years of a small orchard, vegetables can be grown as an intercrop. The use of Biodynamic compost on the vegetables will help build up the soil for the orchard. As the tree roots extend, reduce the vegetable area accordingly; especially avoid disturbing citrus roots. Mulch can be used around the young trees in the first few summers. As the trees grow older, sow the orchard down to a good pasture mix, ensuring that adequate amounts of red and white clover are included. On a larger scale it is best to sow pasture straight after forming the raised rows.

Red and white clover provide most of the nitrogen needs of deciduous fruit trees, though periodic supplementation with small amounts of compost, blood and bone, or pelletised manures may be required. Manures are best used on the pasture when the trees are dormant – or on green manure crops sown in the orchard in autumn, and disced in or slashed when ready, to feed the trees during the following spring.

Citrus trees need large amounts of nitrogen and can safely be (lightly) mulched with chicken manure, best after fruiting. This won't overfeed the tree with nitrogen or cause the bitterness characteristic of conventionally grown lemons. Make sure that no mulch material touches the trunk.

Shelter belts should be planted, if necessary, to shelter the orchard from prevailing winds.

Biodynamic Sprays

Spray prepared 500 every spring and autumn when soil is moist and warm. Slash the pasture before spraying so that droplets will contact the soil. Preparation 501 is used when there is excess moisture, lushness or fertility. Apply it when the leaves are well established in early to mid spring after flowering, when the fruit is set but still very small. It can also be used later, when the fruit is fully grown, to aid ripening and sweetness. If used between these times, it may cause excessive fruit drop.

Excessive Rain

Use casuarina or equisetum tea, alternating with 501. These sprays help strengthen and harden the plant and counter the conditions that lead to fungal problems.

Sudden Cold Snaps in Spring

Valerian spray will bring warmth and restore balance.

Some Pest and Disease Strategies

Black spot and brown rot

Use casuarina or equisetum tea, diluted to tea colour. Stir as for 500 for 20 minutes before applying, and saturate the whole plant including the roots. Other available treatments include wettable sulphur, potassium bicarbonate (e.g. Ecocarb), sodium silicate solution and light copper sprays (200g per 100 litres of water) during a wet spring. Seaweed solution sprayed post-harvest helps the breakdown of the leaves.

Curly leaf and shot hole

As soon as these diseases appear, spray with 0.75% sodium silicate solution (75ml of sodium silicate in 10 litres of water). If the leaves are young, use 0.5% solution (50ml of sodium silicate in 10 litres of water). Remember to wear eye protection. Alternatively, copper (Bordeaux mixture, copper hydroxide or copper oxychloride) may be sprayed – check your national organic standards first – after pruning (if Biodynamic winter spray is not used) and at budswell before any leaves appear.

Scale on citrus

A spray made from kaolin clay and sodium silicate can be used, or cold starch in water. Do not cover the whole tree; spray just the affected areas. Alternatively, lime or wood ash can be blown over the whole tree. White oil is also allowable under the National Standard for Organic and Bio-Dynamic Produce (Australia).

San Jose scale

Black spot treatment generally deals with this adequately. Also: lime sulphur, Biodynamic winter tree spray and later sprays of copper, sodium silicate or casuarina/equisetum if necessary; or winter oil and, later, lime sulphur followed by copper, sodium silicate or casuarina/equisetum.

Phytophthora and *Pythium*

These are destructive soil-borne plant fungal pathogens. *Phytophthora* affects a wide range of species including eucalypts, avocadoes, macadamia, stone fruit and pineapples. *Pythium* mainly affects herbaceous plants. Good drainage is critically important in prevention and treatment. Apply alternate sprays of prepared 500 and casuarina every second day (500 one day, casuarina the next). Repeat three or four times then finish the procedure. Ensure that long grass is removed from the base of the trunk and that adequate space and light are provided. Trees must have very good drainage.

Codling moth

Keep a close eye on your apple and pear trees and remove and destroy any fruit that has brown detritus exuding from a small hole. In early to mid spring, tie a 10-cm-wide corrugated cardboard or hessian strip around the tree trunks. Inspect each month until late autumn and destroy any grubs found. Keep the bark healthy and don't allow loose, rough bark to remain, which would provide ideal conditions for codling moths to pupate. The Biodynamic winter tree spray helps keep the bark smooth and healthy. Pheromone ties (codling moth sex attractant) can be attached to the trees to disrupt breeding. Eggs of *Trichogramma* wasp (a codling moth predator) can be purchased and placed in the trees from the time the moths start hatching. Other strategies include the application of viral diseases of codling moth such as Madex and Cyd-X, which attack the active moths; entomopathic nematodes, which target the larvae during the winter; and *Mastrus* wasps.

Pear and cherry slug

Spray after dark (to avoid killing beneficial insects) with natural pyrethrum (without piperonyl butoxide). Post-harvest, dust the leaves with lime or wood ash or spray with neem extract or diluted liquid lime.

Fruit fly

Pheromone lures attract and kill the female flies. Various traps that contain a liquid lure (some with an enclosed fumigant) attract and kill the flies. Those with chemical fumigants must be disposed of carefully off-farm after use. Molasses-based lures containing a natural toxin (spinosad) can be sprayed on the leaves when the fruit flies are active. Check national organic standards.

Earwigs

Earwigs feed in trees overnight. Apply double-sided yellow sticky tape (about 150mm wide) to the lower trunks after 9am, when the earwigs have returned to the ground, early in the season, when these insects start to become active.

Figure 23.5. Biodynamic bananas, Carnarvon, Western Australia.

Chapter 24
Animal Health

All good livestock farmers are deeply committed to keeping their animals healthy and contented. Not only is this a natural consequence of their genuine love for their animals and concern for their welfare, but it is also essential economically. Organic and Biodynamic farmers, along with many conventional farmers, are committed to free range systems, where animals are allowed to express their natural instincts rather than being confined in unnatural, intensive settings. Consumers are also demanding this, and 'free range' products are becoming more widely available.

Healthy Soils, Healthy Animals

Biodynamic management progressively develops deep, well-structured, friable, fertile soils that are well supplied with colloidal humus and relatively free of dissolved elements. Plants that grow on such soils are healthy, upright and vibrant. Elements taken up by the plant roots are fully assimilated in the plant tissues, resulting in high levels of quality vitamins, minerals, proteins, etc., providing balanced nutrition to animals and humans alike and building robust health and resistance to disease.

By contrast, conventional agriculture, which applies water-soluble artificial fertilisers, results in soils that are less well structured, with less humus, soils that feed plants indiscriminately with elements dissolved in the soil water. Plants that grow in such conditions are metabolically unbalanced, the plant cells being 'blown up' with excess unassimilated mineral salts. Unassimilated minerals such as nitrates can cause serious harm to animal (and human) health and result in poorly flavoured produce. To make matters worse, the overblown plants are very susceptible to attack by a myriad of plant diseases and insect pests and must be protected with chemical sprays, which further compromises the health of both animal and human consumers.

Instinct tells animals to avoid such unbalanced, poorly flavoured produce unless they have no choice, and should ring alarm bells for people too if we are at all awake. Try one simple test: buy a conventional banana and a Biodynamic or organic banana and do a blind taste test. There is generally a vast difference in flavour, and more so with Biodynamic fruit.

The above comparison really tells the story of animal (and human) health: Biodynamics (and

Figure 24.1. Biodynamic White Suffolk ewes and lambs, Oberon, New South Wales. Photo: Tam Kurtz.

also good organic practice) builds robust animal health. This is borne out by the many examples given in this book of Biodynamic farmers' lived experience with animal health, and – more specifically – by the veterinary results of the 1991 comparisons of ten matched pairs of Biodynamic and conventional dairy farms in Victoria (see p. 22–23), which clearly demonstrated superior health, fertility and longevity in Biodynamic dairy cattle compared with cattle on the conventional farms.

Natural Feeding

Of course, animals on Biodynamic farms are fed Biodynamic food (including the so-important pasture), which is highly nutritious, full of vitality and relatively free of chemical pesticide, herbicide and fungicide residues.[1] This builds robust health and resistance to disease, as it does in humans.

Under organic/Biodynamic certification standards, bought-in feed must also be certified Biodynamic, apart from mineral supplements. Where certified feed supplements are not available, a small amount of uncertified feed is allowed, provided it is free of prohibited substances or contaminants. In Australia this can be up to 5% of the total feed ration; the percentage varies from country to country. Natural feeding includes the recognition that some foods are not naturally part of an animal's diet. No meat product should ever be fed to a herbivore. Sheep brains included in bovine feed in Europe caused BSE 'mad cow disease' – and Rudolf Steiner had said in 1924 'if you feed the ox meat, the ox will go mad'. Grain feeding of dairy cows should be minimal. Biodynamic dairy farmers never feed more than 4kg per head per day, and most don't go over 2kg per day, compared with up to 12kg per day on some conventional dairy farms. At this lower rate of grain feeding, cows are much healthier and have longer productive lives and the milk is of much higher quality.

Mineral Supplementation

Biodynamic pastures and crops, with their deeply penetrating, active feeder roots, bring high levels

Figure 24.2. Biodynamic pigs and piglets, Goomalling, Western Australia.

of nutrition into pastures and grains and other crops. In many areas, varying to some extent from season to season, mineral deficiencies exist in the soil and plants. For instance, coastal soils in Australia are often deficient in copper, selenium and cobalt. Animals are very capable of balancing their diet, seeking exactly what they need. In areas with soil mineral deficiencies, farmers should supply mineral licks or blocks for livestock so they can balance their diet. These can be ready-formulated blocks or mixes, suited to the particular deficiencies common in particular areas, or the ingredients can be supplied in separate containers.

Supplying individual minerals separately allows animals to accurately select what they need. The weakness of ready-mixed formulations is that animals may have to consume a large amount to remedy a deficiency of one element, whereas with self-selection there is no waste.

Some caution is required with copper sulphate and sulphur provided alone. Sulphur provided by itself should be prilled (granulated), since powdered sulphur can be inhaled, causing lung problems.

John and Bernadette Cashmore, retired Biodynamic grain and sheep producers in Western Australia, provided minerals for their sheep in separate containers in small, weatherproof shelters in each paddock. The minerals provided were copper sulphate, prilled sulphur, food grade phosphate, sodium bicarbonate, gypsum, lime, starch, salt, sugar and apple cider vinegar. Some of the minerals were combined with the sugar but were also provided individually to give adequate choice. The 250kg of food grade phosphate provided to the sheep in this way over a 6-year period was the only phosphate applied to the farm during that time.

Some common mineral licks

- Mineral mixes can include dolomite, seaweed meal, yellow dusting sulphur, copper sulphate, cod liver oil, apple cider vinegar, salt, and trace minerals that are known to be lacking.
- Australian animal nutrition expert Pat Coleby suggests a 'Basic Stock Lick' comprising a mixture of 25kg of dolomite, 4 kg of copper sulphate, 4kg of 99% yellow dusting sulphur (as used in orchards) and 4kg of seaweed meal.[2] Many farmers have salt available separately.
- For cattle and sheep, mix eight parts dolomite to one part each of copper sulphate, kelp powder and sulphur.
- Sulphur licks strengthen animals against lice.
- Dolomite and sulphur in licks help protect against fly strike.
- Salt 63%, sulphur 20%, limestone 15% and copper sulphate 2%.
- Some farmers, in areas where soils are generally well supplied with minerals, simply provide salt or a salt lick block, often sea salt.

Natural Behaviour

Good farmers endeavour to provide as natural and stress-free an environment as possible for their animals, not only because of their genuine concern and love for their animals, but also because a contented animal thrives, grows and produces at an optimal level, which positively affects the farmer's income. This includes the provision of appropriate feed and water, shelter from climatic extremes, protection from predators, physical handling in a way that minimises stress or pain, rapid diagnosis and treatment of illness or injury, freedom to exercise and move naturally, social contact with other animals and the opportunity to display appropriate natural behaviours.

Although the trend is very much towards keeping animals under free range conditions, many are still confined and overcrowded. The official definition of 'free range' for poultry in Australia is a maximum of 10,000 birds per hectare. Some producers use this to mislead consumers about the life enjoyed by their birds. Genuine free range as practised by organic and Biodynamic producers involves far lower stocking rates, ensuring that birds always have access to healthy pasture, not just dust or mud.

Rotational Grazing

Rotational grazing is a key factor in animal health, being excellent for root and soil development as well as for worm control. It has been an essential feature of Biodynamic farming in Australia since the 1950s. Smaller paddocks (large paddocks can be subdivided with electric strip grazing) and frequent moves are best. Grass is best grazed to 40–60mm, no lower. This allows faster regrowth.

Breeding for Health

When animals are routinely treated with veterinary medicines, particularly regular worming, innate weaknesses are masked. Certified Biodynamic (and organic) farmers are restricted in their use of veterinary drugs. Permitted materials for livestock pest and disease control are listed in Appendix D of the Australian

National Standard for Organic and Bio-Dynamic Produce (and the equivalent section of the organic standards of other countries).

Certified farmers may use veterinary drugs when allowable treatments have failed or if the situation requires immediate veterinary intervention. In such situations, the treated animal must be quarantined in a designated farm quarantine area for three times the withholding period of the drug or 3 weeks, whichever is longer, and can never be sold as certified. Some animal products from these animals can be sold as certified after minimum management periods. For instance, milk can again be certified 180 days after treatment with a veterinary drug (check your country's organic/ Biodynamic standards).

In the absence of regular chemical drenching, stock that are susceptible to worms are more easily identified by the farmer. For instance, dirty bottoms in sheep are a simple indicator, bearing in mind that this can be a more general occurrence in a very wet period. These animals can be treated with natural, allowable materials. If the affected animals show a continuing susceptibility to worms, they are not used for breeding and are usually sold. If an animal is sick enough to require chemical drenching and the farmer decides not to keep it, it is quarantined for the period required on the label and then sold (uncertified). Some sheep can be badly affected by worms and lose production without showing a dirty bottom. It is important to be able to identify these animals; careful observation is critical.

Testing of ram faeces has been shown to be a very useful long-term strategy to improve worm resistance in a flock. Dayle and Terri Lloyd, Western Australian Biodynamic grain/sheep farmers, running 3500 Merinos, ran a testing programme with their rams for some time. Worm problems are easier to identify with rams than with ewes, owing to varying hormone levels in ewes, and each ram, serving 50 ewes, has a major influence on the flock. This programme has refined selection for worm resistance in the flock. In wet years, the Lloyds' sheep now cope very well without treatment, whereas neighbouring conventional farmers have ongoing problems in spite of drenching. The consequence of not masking weaknesses by regularly treating animals with veterinary drugs such as worm drenches is that stock on Biodynamic farms are specifically bred for health and resistance to disease and become progressively more robust with each succeeding generation.

Vaccination

Certified farmers in many countries are allowed to vaccinate animals where it can be shown that other management practices are insufficient to protect against disease or illness, provided the vaccines are free of genetically modified organisms. This normally applies where particular diseases are prevalent in an area and pose a high risk to stock. We do not view vaccination, per se, as problematic, in fact in such areas, it is definitely recommended to safeguard livestock. Many Biodynamic farmers don't need to vaccinate their animals.

Some Specific Treatments

The following treatments have been successfully used by Biodynamic farmers to treat health problems in their animals. Many farmers also have success with homeopathic remedies, available from various suppliers. This is not an exhaustive list of treatments but rather an indicator of the types of natural treatments available. Some farmers also use the natural treatments detailed by Juliette de Bairacli Levy in *The Complete Herbal Handbook for Farm and Stable*.[3] If any animal fails to respond to such natural treatments, recourse to conventional treatments may be needed in order to restore health or save the animal. Such animals need to be quarantined according to certification standards and may need to be sold uncertified after any required withholding period.

Sheep and goats

Worms

To some extent, the presence of worms is natural and cyclic and should not particularly affect healthy animals. Provision of healthy, balanced feed (including adequate copper), rotational grazing and selective breeding will all help overcome worm problems in time. Well-organised rotational grazing is one of the most effective worm control methods. If worms do become a problem, natural drenches can be used. Drenching is generally done at full moon, when worms are most active. Some of the worm treatments used by Biodynamic farmers include:

- Apple cider vinegar, 13ml per head.
- Apple cider vinegar and garlic drench.
- 20 litres of apple cider vinegar, a handful of crushed pumpkin seeds, a little kelp powder, a little wormwood – mix and allow to ferment for a week, then pour on hay.
- Seaweed, cod liver oil and apple cider vinegar drench.
- 25ml of neem oil – for weaners and ewes.
- Nutrimol – a seaweed-based drench.
- Seaweed, apple cider vinegar and garlic powder at 10ml per lamb or 2ml per 10kg live weight.
- Drench lambs with 1.8ml per 10 kg live weight of a solution of 80g copper sulphate per litre of water.
- Mix 250ml of vegetable oil, 250ml of apple cider vinegar, 250ml of Nutrimol, 250ml of water, 1 teaspoon of garlic, 50g of copper sulphate and apply to goats and ewes at 3ml per 10kg live weight.

General tonic for sheep and lambs
TNN Mineral-Plus. (TNN Australia also offers a wide range of trace elements for stock supplementation.)

Sick lamb drench
Equal parts apple cider vinegar, cod liver oil and seaweed concentrate (e.g. Nutrimol or Vitec). Use 20ml per head each day for 3 days while fasting them. Vitec Organics produce a range of natural animal supplements and natural treatments.

Lice
- Provide sulphur licks.
- Spinosad-based dips or sprays, such as Extinosad (an approved organic input in Australia). Spinosad is a natural substance made by a soil bacterium.
- Jet (spray externally) with a mix of 150 litres of water, 1 litre of neem oil and 1 litre of

eucalyptus oil, in the evening, before dark.
- Wide boundary shelter belts to prevent contact with neighbours' sheep.

Fly strike
Generally, Biodynamic sheep resist fly strike well and their manure is less smelly than that of conventional manure and thus less attractive to flies.

- Apply a mix of eucalyptus oil, citronella and natural detergent such as potassium soap (e.g. Natrasoap).
- Apply citronella and apple cider vinegar.
- Natural pyrethrum (without piperonyl butoxide) and garlic, applied externally.
- Natural pyrethrum (without piperonyl butoxide) and eucalyptus oil mix, applied externally.
- External pyrethrum.
- Sulphur, copper, apple cider vinegar, and borax mix for fly dressing.
- Pyrethrum and mud mix for dressing.
- Pyrethrum and clay water for dressing.
- A dash of eucalyptus oil in a spray bottle of apple cider vinegar.
- Eucalyptus oil applied externally.

Cattle

General tonic for cattle
TNN Mineral-Plus.

Worms
See introductory comments for worms in sheep on p.185. Use:
- Garlic and apple cider vinegar, 30–45ml

per head.
- Homeopathic sulphur drops.

Lice
- Apply sulphur dust along backline.
- Spray cattle with a spinosad-based product such as Extinosad (see p. 185) if allowed by your country's organic/Biodynamic standards.
- Mix 400ml of eucalyptus oil, 400ml of coconut oil and 50 litres of hot water and allow the mixture to cool to a safe temperature before applying to the animal's back.

Ticks
- Neem extract and vegetable oil on the back rubber that dairy cows rub their back on as they enter the dairy. Similar arrangements can be made for beef cattle.
- Homeopathic neem powder.
- Other homeopathic remedies – one tropical dairy farmer uses a homeopathic tick remedy at morning milking and a tick pepper (see Chapter 22) put in the trough at evening milking.
- Neem oil.
- Newly introduced cows can require a year before health improves and ticks become less of a problem.

Buffalo fly
- Neem oil as back rub.
- Natural pyrethrum spray (without piperonyl butoxide) on cows every fortnight.
- Organica Cattle Coat, a natural back rub oil (containing essential oils) for use with back rubs and which can also be applied via post

rubbers or sprayed on.

- Fly traps.

Liver fluke

One teaspoon of copper sulphate in stock troughs when needed.

Flies bothering cows at milking

Spray cows' legs with water when they come into the dairy, and let them stand for a few minutes before milking. Try mixing apple cider vinegar and tea tree oil and spraying the legs. Keep the dairy clean of manure during milking. Spread strong-smelling herbs on the dairy floor. Large fans in the dairy are a great help.

Mastitis

- Japanese mint rubbed into the udder.
- Homeopathic treatment with goldenseal tincture for subclinical mastitis.
- Homeopathic ABC (aconite, belladonna, chamomile).

Teat spray

Use 2 litres of glycerine and 50ml of tea tree oil in 25 litres of water. Conventional iodine teat sprays are also allowed under many countries' organic standards.

Milk fever

A calcium and magnesium injection is allowable under many certification standards.

Scours in calves

Homeopathic salmonella nosode.

Open wounds

Powdered gentian, applied externally.

Pink eye

- Draw 15ml of cod liver oil in a syringe. Apply 3ml to each eye and the rest by mouth. Also put some high-quality feed or hay in a feed trough and pour cod liver oil over it.
- Use 20ml of cod liver oil by mouth and a squirt of golden seal solution (six drops of high-quality Golden Seal tincture such as that from Herbal Extract Company in 300ml water) in the eye.
- Coconut oil in the eye.

Foot rot

- Copper sulphate taped to the foot, combined with homeopathic drops.
- Hydrogen peroxide applied to the foot.

Poultry

Lice

- Dust with dolomite or diatomaceous earth. Wear a face mask while applying diatomaceous earth to avoid breathing in the dust.
- Provide sulphur in feed.
- Dust with derris or sulphur or a combination as in Pestene. Also add this to dust baths in sunny spots.
- Treat sheds and perches by painting them with vegetable oil.

Scaly leg

Mix neem oil and diatomaceous earth and apply the mixture to the legs.

Figure 24.3. Young Biodynamic Murray Grey cattle, Burrell Creek, New South Wales.

Pigs

- Use garlic for worms in adult pigs.
- Use citrus seed extract (e.g. from grapefruit) for worms in piglets.

FOOTNOTES FOR CHAPTER 24
1. Owing to the widespread contamination of the environment by chemical use on conventional farms, complete absence of residues cannot be guaranteed even on certified farms.
2. Pat Coleby, *Natural Farming*, Scribe, Brunswick, Victoria, 2004, p. 168.
3. Juliette de Bairacli Levy, *The Complete Herbal Handbook for Farm and Stable*, Faber & Faber, London, 1952.

Part 3

Biodynamic Farm Studies

Case studies of some Australian Demeter method

Biodynamic farms in Australia and elsewhere

Chapter 25

Biodynamic Dairy Farming, Nathalia, Victoria, Australia

It is not hard to find Mark and Lynne Peterson's irrigation dairy farm near Nathalia in Victoria. With tens of thousands of trees planted over the last 30 years, it stands out as a lush green oasis in the midst of farms that are relatively bare. The atmosphere created is one of life, fertility and lushness – surprising in a district whose long-term annual rainfall is only around 375 mm.

Mark and Lynne supplied Demeter certified Biodynamic milk (along with several other Victorian Biodynamic dairy farmers) to Parmalat for over a decade. The milk was distributed widely in supermarkets until a long drought reduced supply and led to Parmalat ceasing to market it as Biodynamic and instead including it with their organic milk. Consumers were upset at losing their better-tasting Biodynamic milk and clamoured for a return of pure Biodynamic milk.

After much research and discussion with Peter Podolinsky at the Biodynamic Marketing Company (BDMC), Mark and Lynne found a solution: the BDMC agreed to buy their milk at the same price as Parmalat were paying, pay a tanker to pick it up and transport it to a dairy factory in Kyabram and pay the factory to pasteurise and bottle it. The BDMC would

Figure 25.1. Mark Peterson.

then distribute the milk through health food shops in Victoria. This was a risky project for the BDMC, since they had to sell over 6,000 litres a week to break even, but the company is there to assist farmers and consumers and often takes a long-term view to facilitate Biodynamic produce reaching the consumer as efficiently as possible.

Mark's father and mother bought the original 123-acre (50-hectare) irrigation farm in 1976. With several extra purchases the farm had reached 350 acres (142 hectares) by 2010. During the 1983 drought, when times were very hard, Mark's father bought a backhoe business and Mark became the share farmer. A few years before, he had completed a farm apprenticeship through Shepparton TAFE (Technical and Further Education). In the third year of his apprenticeship, the instructor organised a visit to a Biodynamic dairy farm run by Maurie and Nance Fenell at Cogupna. He told the students the Fenells didn't have any trouble with bloat, didn't have to drench their cows and didn't have much mastitis. The other apprentices scoffed and said that this was not possible, but Mark was intrigued and asked Maurie many questions. Maurie was milking 60 cows on 50 acres (20 hectares), with irrigation, and rearing replacement milkers, only buying in a bit of hay and small amounts of grain. His cows looked magnificent; his pastures were lush and dense. This greatly impressed Mark, along with the fact that the farm was a more or less self-sufficient, low-cost operation involving no chemicals and virtually no fertiliser inputs. Later, in the mid-1980s, Mark and his dad watched *A Winter's Tale*, the ABC TV feature on Alex Podolinsky and Australian Demeter Biodynamic farming.

Dairy farming margins for conventional farmers were very poor at the time, as they are now, and Mark could see they were getting nowhere. He said he wanted to start Biodynamic farming, which seemed like a low-cost, healthy alternative, and his father agreed. Maurie Fenell recommended he contact Trevor Cobbledick for guidance.[1] Mark

started spraying 500 in 1987, having found a second-hand stirring machine and spray-out rig. The effects weren't instant, but crept up on them slowly. They made an initial mistake by stopping drenching their cows and calves too soon. The calves looked terrible after a while and Mark asked Trevor for advice. Trevor told him that you can't just go cold turkey; you have to wean the animals off chemical drenches slowly. Trevor advised Mark to keep the animals moving through a good pasture rotation. Mark drenched the calves and they got through okay. The milkers didn't need drenching, however.

After 3 years of Biodynamics, during which Mark and Lynne bought the farm, the calves, which looked very healthy, were tested for worms. The reading was so high that the vet rang and said Mark should drench them immediately. He asked Trevor what he should do. Trevor said, 'What would you have done if you hadn't seen the test results?' Mark said, 'Nothing.' Trevor said, 'Well, do nothing and see what happens.' So Mark did nothing. A few months later, after another test, the vet said, 'So you drenched those calves?' Mark said, 'No.' The vet was amazed, since all the calves, except one or two, had almost zero worm counts.

After 35 years of Biodynamics, Mark and Lynne's soil has changed from a red loamy soil only a few inches deep, with clay underneath, to an almost black topsoil which extends to 450mm depth. This has puzzled earthmoving contractors and machinery salesmen, who know the area's soils intimately and can't understand why this soil is so much better.

Biodynamic Preparations

Mark sprays prepared 500 (stirred in a 20-acre [8-hectare] stirring machine) with a Bill Chandler reciprocating sprayer (see Figures 11.19–11.21) twice a year on the perennial pastures (spring and autumn) and once a year (autumn) on the annual pastures. Summer crops are also sprayed with 500. Mark irrigates, and then, when he can drive on the field and the tractor tyre cleats are just making an impression, it is time to spray. If the cleats are sinking in, it's too soon. If they make no impression, it's too late (too dry). Preparation 501 is never needed in this low-rainfall, sunny climate.

Pastures

The Petersons maintain a mix of perennial (plants grow, set seed, go dormant and, later, regrow) and annual (plants grow, set seed and die) pastures. During the long drought, annual pastures had to predominate because of the difficulty of keeping perennial pastures going with inadequate water allocations. The main grass species is rye grass. Perennial pastures comprise rye grass, white clover, paspalum, chicory, plaintain and many other lesser but nonetheless important species. Annual pastures include sub clover, Shaftal Persian clover, winter-active fescues and others, again on a rye grass base.

Mark is very impressed with modern winter-active fescues. They are tricky to establish, requiring a fine seed bed, no competition while they are young, and careful grazing, but once established they persist very well and are very hardy. They are also more palatable for dairy cows than older, rougher-leaved varieties. Once established, and over-

Figure 25.2. Plantain that came up after summer rain – glow-green, upright, full of life and relished by the cows.

Figure 25.3. Fescue and clover.

sown with sub clover, the pasture will last for 15 years with careful management and no overgrazing.

Over-sowing existing pastures is used extensively to re-establish species after drought or to rebalance pasture composition. Perennial pastures are over-sown every 3 years or so, mainly with rye grass.

Mark has learnt that ploughing a field using the Rehabilitator plough does leave the ground

Figure 25.4. Newly sown paddock with new irrigation channel.

quite soft and that it is often best to grow a cereal crop in the first year before re-sowing to pasture and grazing. The sowing of fescue requires a fine seed bed on top; you drop the tiny fescue seed on top, press it in with (press wheel) tyres and get it well established before grazing.

Provided enough irrigation water is available, Mark plans to start green manuring a field or two each year to intensively improve the soil.

Pastures are harrowed after the cows move on, to spread the manure evenly. Mark uses a very wide set of pasture harrows and can do a field in 15 minutes or so. The best time is in the morning while the dew is still on the grass.

Weeds are generally not a problem. Some species that are commonly regarded as weeds are relished by the cattle and are highly nutritious under Biodynamics. A few thistles come and go. On the Petersons' farm, they are opportunistic weeds that come when soil is disturbed and that don't like irrigation or competition. They do their job and then disappear. Paterson's curse (toxic to cattle) is

hand pulled if it appears anywhere. Bindiis (prickly weeds) come in on trucks and milk tankers, but again they don't like competition. Manure harrowing plays a very important part in weed management, since cow pats left in place acidify the soil and weeds come to balance the problem.

Irrigation

Irrigation water comes from the Murray irrigation scheme. Each field is laser levelled, and flood irrigated through timer-controlled gates. After leaving the field, the water returns to a recycling dam from where it is pumped back on

Figure 25.5. New irrigation channel.

Figure 25.6. Water being pumped from recycling dam to irrigate paddocks.

Figure 25.7. Paddock being flood irrigated.

Figure 25.8. Harrowing pasture to spread manure after grazing. Photo: Mark Peterson.

Figure 25.9. Heifers.

to other fields. During the drought, allocations fell as low as 35%, but, with full reservoirs following a series of high-rainfall years, things are currently looking much better. Under current rules for water trading, farmers can buy water and carry it over to the next year, which gives them much more control of their situation. When the water is available, Mark irrigates every 10–14 days (conventional farmers irrigate every 7 days). However, during the drought the subsoil dried out so much that he had to come back to every 9 or 10 days.

Figure 25.10. Milkers. Photo: Mark Peterson.

Figure 25.12. Manure pusher – used with quad bike.

Figure 25.11. Dairy.

Rotational Grazing

For over 70 years, Podolinsky stressed the importance of good rotational grazing and strip grazing. Mark and Lynne's rotational grazing set-up evolved over time. On the basis of Maurie's advice, together with their own experience, they now have the young stock, the dry cows and the milkers on three separate rotations, in three separate areas. Maurie's set-up, on his tiny, triangular-shaped 50-acre (20 hectare) farm, reminded Mark of three cogs going around together. Under this system, each group of cattle gets fresh grass and doesn't have to follow another group, causing worm problems, while also eating the best grass species lower than ideal while leaving the not-so-good grass.

Mark and Lynne have a total of 70 fields, many of which are around 7 acres (3 hectares) or even less in area. Each field is strip grazed with electric fences. Mark has two electric fences in the paddock, so that he can just drop one when the cows need to be moved, and they can run straight into the next area. With strip grazing he can get almost twice the feed out of a paddock and has so much more control than if the cows were given the whole paddock at once.

Mark took a pasture feeding course on which he learnt a lot about the optimum stages of grass growth. Optimum height after grazing is 4–6cm. If the cows start to wander back and eat the regrowth (back-grazing), you have left them there too long. Back-grazing sets the grass back badly. Optimum height before grazing varies with the time of year and the species involved. The predominant grass, rye grass,

should ideally be at the three-leaf stage. Once the fourth leaf starts to grow, the first leaf dies. However, this doesn't apply in spring, when the grass grows so much more quickly and can have five or six leaves.

Mark doesn't look too far ahead of the milkers. To determine how the rotation is going, he looks at where they are going tomorrow and where they were yesterday. If you are moving them too quickly, you can cut the paddocks into four instead of three, add more hay or silage, or, for instance, graze pastures until 11am and then put them in a non-pasture feeding area with hay and silage. You have to get the balance right between what's good for the pasture and what's good-quality feed for milk production. In spring the rotation may be down to 20 days, and in late summer, when the paspalum takes off, the rotation may go down to 15 days. Paspalum mustn't be allowed to get away too much. For one thing, it makes over-sowing very difficult. If there are summer rains, Mark sometimes has to top the fields and rake the paspalum off to the sides lest it block the sowing machine. In winter, the rotation may go out to 90 or 100 days. Mark has a very good computer program to help with planning rotational grazing: he enters a rating for each paddock (poor, good, excellent, etc.) and the program calculates how many feeds you will get from it. The program depends very much on the farmer's observation and assessment of the pasture. It was a good training tool for a few years, but now Mark knows instinctively how to manage the rotation. However, the program will still be useful when Mark and Lynne go away, enabling their relief worker to accurately manage the rotation.

Fertiliser Inputs

Mark applied no fertiliser for 20 years, but, because of the long years of drought, he has for the last few years applied 20kg per acre (50kg per hectare) of guano with the seed when sowing annual pastures, just to give the seedlings enough of a start to get going. He doesn't use any lime; soil tests always come back with a good pH reading.

Compost

Mark makes and spreads 120–150 tonnes of Biodynamic compost every year. After milking, he pushes the manure off the yard into a holding pit, using a rubber scraper that fits on to his quad bike. A little straw is thrown over the manure each day to protect it from drying out. Every week, he empties the pit and takes the manure to the compost area. When enough material has built up, he will make a large Biodynamic compost heap. He makes three or four heaps each year, spreading the compost at the rate of 5–8m³ per acre (12–20m³ per hectare).

Mark turns the material several times while it is building up, by pushing it over with the front-end loader. He keeps it at no more than 60cm in height to prevent it heating up before the heap is built. By the time he is ready to build the heap, the material is already well mixed. He uses a Krone manure spreader with a metal cowling over the beater blades to build the heap, moving forward progressively as it comes up to height.

Mark monitors the temperature carefully, trying to avoid temperatures over 45°C. Only when the temperature drops to 35°C does he order the Biodynamic compost preparations. He makes sure that the preps are inserted in the heap within a few hours of their arrival in the mail. Even after the preps are in, Mark only covers the heap with straw or old hay in stages from the bottom up. He feels that, this way, heat can escape more easily. Only when he is satisfied with the temperature will he finally cover the whole heap. In this low-rainfall climate, he leaves a 60-cm-wide flat top on the heap to allow rain to get in.

When deciding when to spread the compost, many factors have to be considered. What will the weather do? What's the grass cover? When will the next irrigation water come? If a compromise has to be made, Mark prefers to spread when the compost is a few weeks off being mature rather than a few weeks after it matures. He has noticed that, even if there are still some green bits in the heap, by the time it has been trucked to the intended field and dumped in a heap for a few days, the extra aeration will have finished it off. Mark drives the manure spreader up and down the field, with the metal cowling removed, and on each run a worker drops a front-end loader bucket in. This way, the spreader is not put under undue strain from carting a very full load to the paddock and spreading it, and the weight on the pasture is minimised. The pasture should have a good grass cover (100–150mm), the compost is spread in the late afternoon, and the field is irrigated soon after spreading. Good grass cover helps shade the compost, and irrigation helps it incorporate with the soil.

Mark arranges things so that, straight after spreading one compost heap, he builds the next one, while the spreader is still dirty. After building the heap, he hoses the Krone down thoroughly. For longevity, it should be shedded.

Trees

The trees planted are all natives. The first plantation was of eucalypts from northern New South Wales, which suited the warm climate and irrigation water, but when water became scarce during the drought they struggled. Later plantings were of trees from the local area that suited the superb parrot, a beautiful but endangered species. The Petersons are hoping to attract these parrots to their property and thereby help establish links to the Barmah State Forest habitat. They now tend to plant a higher percentage of understorey species and fewer eucalypts. In drought times the larger eucalypts compete with the pastures, and some plantings have had to be thinned. Not only do the tree plantings provide shelter for stock and pastures, and bring nutrients up from deep in the subsoil, but they also provide all the Petersons' firewood requirements and they can also sell firewood.

The Herd

Mark and Lynne milk a mixed herd of mainly Jerseys and Friesians, resulting in milk with about 5% butterfat. Mark has increased the percentage of Friesians to reduce the butterfat levels, since some people don't like their milk to be too creamy. A low-fat milk is also produced at the dairy factory.

When Maurie and Nance Fenell retired they helped the Petersons purchase their Jersey stud herd, since they wanted the cows to go to a good home. The Petersons' predominantly Friesian herd was culled heavily in order to introduce the new cows, which, being from a Demeter certified property, could fit straight into the milking programme.

Because Mark and Lynne's is one of only two farms currently supplying Demeter certified Biodynamic milk for Victorian consumers, the pressure is really on to produce the highest quality possible, with very even production levels. They ran three calvings per year for some time – in March/April, July/August and November/December – and have now extended that to four per year.

Breeding is three-quarters AI (artificial insemination) and one-quarter natural. The Petersons are using mainly A2 bulls, since many consumers are now asking if their milk is A2.[2] The milk is now predominantly A2. They choose bulls for medium stature and ease of calving rather than high production. Mark generally has to help only one cow per calving group – generally with a turned foot or other simple problem. Calves for replacements come from the autumn and winter calvings only, since summer calf rearing is too difficult in the hot climate.

Apart from grass, hay and silage, Mark and Lynne feed cereal hay and a little grain (all certified Biodynamic). Mark likes grain as a way of getting the cows into the shed and as a supplement to their diet, but not in large quantities. In winter he feeds a couple of kilos per cow per day (up to 4kg in the drought when feed was very scarce), but when grass is growing well he cuts back to 1kg. All feed

Figure 25.13. Well-treed laneway.

that is bought in is from Demeter certified Biodynamic farmers. During the drought, when feed became very scarce and expensive, Mark found that it was more economical to buy water to grow more fresh grass.

The cows and calves are remarkably healthy and free of problems. Drenching is not necessary and antibiotics are never used. Over the years, animals that have problems have been culled and so a strong herd has been selected. Cows are almost never brought in from outside the system. Heifers occasionally are; they become certified after 6 months on the property, before their first calving. Bloat is very rare, despite the lack of preventative treatments. Mark has lost only two cows to bloat in 20

Figure 25.14. New parcel of land that was used for young stock until it reached Demeter standard.

years. Milk fever is uncommon and getting rarer and rarer. Mark takes the cows off pasture a few weeks before calving and gives them plenty of cereal hay and a little bit of grain.

If milk fever occurs, it is usually in older cows and Mark is able (under certification standards) to use a calcium injection to cure it. Cows are dried off by simply stopping milking. Dry cow treatments are never used.

Mark has to watch somatic cell count carefully to maintain the high standard and unusually long shelf life of the milk. Udders are not washed unless dirty. Small amounts of dirt are wiped off with paper towel. If a cow has gone down in the yard, the udder is washed and usually dries off by itself in summer. In winter, paper towel is used to dry the udder. Teat spray used to be made up with 40ml of Biodynamic tea tree oil in 25 litres of water with 3 litres of glycerine (glycerine helps soften the teats), but Mark now uses iodine-based teat spray. The cell count is usually around 200 (industry standard is 250) and very little mastitis occurs. When it does occur (usually in older cows), in a late-lactation cow, she will be dried off. An older cow may be culled straight away (after treatment). Younger cows will generally fully recover. An older cow may subsequently have a dry quarter and be milked as a 'three titter', but if she gets another bout of mastitis she will be culled.

Sometimes a cow will develop a sore foot. Mark first cleans it out with a brush and water. In one case in ten a lodged stone has caused the problem, but if there is an infection he presses copper sulphate into the foot and then tapes between the claws a kitchen cloth packed with copper sulphate and rolled into a cigar shape. He tapes diagonally from the claw to the top of the foot – not around the ankle, since that restricts blood flow. The dressing is removed after a few days. Bad cases are kept in a small paddock near the yard to save them walking for a few days. Foot problems can be associated with feeding higher levels of wheat. Barley and triticale don't cause problems, and if Mark has to use wheat he will combine it with other grains. As to fertility, very few cows fail to get in calf and Mark doesn't have to cull for fertility issues. Fertility is partly related to the higher levels of selenium in Biodynamic cows (see p. 22–23). With four calvings a year, a missed pregnancy can soon be rectified.

Mark and Lynne are currently milking 200 cows (as well as carrying replacements and growing most of their hay), producing 1.1 million litres per year.

Milk Production

The cows are milked in a 12-a-side herringbone dairy and produce 5,000–5,500 litres each per year. This is relatively low compared with cows on conventional farms, but the Petersons' cows are smaller-bodied and cost much less to feed than the very large cows on many conventional farms. They also have a much longer productive life than conventionally farmed cows. The milk is also of much higher quality. It was awarded an exceptionally long shelf life by the dairy factory and won *The Age* (newspaper) Epicure section blind taste test in which a panel of expert tasters compared milks available on the Victorian market. It was the only milk to be awarded the rating of 'outstanding'. Many consumers have reported that it is the only milk they and their children can drink without suffering from bronchial and ear infections.

On average the Petersons send 22,000 litres of milk per week to the dairy factory, most of which is marketed as Biodynamic. They are paid a substantial premium for their milk by the BDMC.

FOOTNOTES FOR CHAPTER 25
1. Biodynamic cattle and grain grower in Nathalia, Victoria. Mark also got a lot of valuable advice early on from Biodynamic dairy farmers Don Rathbone and Evan Hardy.
2. Milk today contains more than one type of beta casein protein, but originally all cows produced only A2 beta casein milk. There is some evidence that A2 milk may be more digestible for many people.

Chapter 26

Biodynamic Orchard, Merrigum, Victoria, Australia

Greenwood Orchards at Merrigum in the Goulburn Valley, 160km north of Melbourne, is probably the only orchard in Australia that has never used a single application of artificial fertiliser. Lynton Greenwood's grandfather Hal established the orchard in 1906 on what had been a dryland farm. When Lynton's father, Farry, took over, orchardists in the area were still using only organic fertilisers. Lynton remembers, as a boy, helping Farry spread poultry manure by horse and cart at the rate of one shovelful per tree.

In the 1930s, the Victorian Agriculture Department was still advising orchardists to use traditional, proven soil management techniques such as green manuring. As artificial fertilisers became the norm, Farry resisted the trend, sticking with the proven annual application of poultry manure together with green manuring on the 142-acre (57-hectare) orchard. The main markets for their apples (42 acres [17 hectares]) and pears (100 acres [40 hectares]) were European exports and the Sydney wholesale markets. Some chemical applications were required to keep fruit clean and kill codling moth, a requirement of the markets. The organic market didn't develop to any extent in Australia until the mid-1980s.

Figure 26.1. Lynton Greenwood.

In 1964 Farry converted to Biodynamics, under the guidance of Alex Podolinsky. Apart from the usual improvement in soil structure, humus levels and colour, the Greenwoods noticed that bird's-foot trefoil (a valuable legume), which had not been very obvious for years, came back strongly after the first application of 500.

In the early years, Demeter Biodynamic certification (the first organic certification in Australia) catered for growers who were required to spray some chemicals for market requirements, by providing a Grade C certification in addition to Grade A (fully Biodynamic) and Grade B (Biodynamic in conversion). Grade C was later withdrawn as the organic industry and the organic market developed. Until then, Farry used limited chemicals at half recommended rates. Farry became a leading Biodynamic practitioner and a Biodynamic adviser of enormous influence far and wide. He was a creative innovator. After the Second World War, Australia was booming and materials and equipment were very difficult to get. In the 1950s Farry built many pieces of equipment, including the heavy disc plough still used on the orchard today and an electric forklift. He built a mudbrick cool store with 900-mm thick walls, and a huge shed supported by concrete columns, using 44-imperial-gallon drums to contain the concrete.

Farry's wife Audrey was an active participant in the life of the orchard until her late eighties. She died in 2008 at the age of 90, still sharp as a tack. She had a profound influence on many people's lives far and wide and was an immensely encouraging influence on the development of the BDAAA. She was warmly known as the 'grandmother' of the BDAAA.

After Farry's death in 1986, Lynton and his wife Joan took over the running of the orchard. They immediately stopped spraying chemicals on the 40-acre (16-hectare) block nearest the house and progressively stopped spraying the rest of the orchard in the following years. This courageous decision resulted in the loss of the Sydney markets. An inspector found a single codling moth in a pear and Greenwoods was banned. A few difficult years followed, with much hard work in marketing the Biodynamic fruit in a still-fledgling organic market. Of great help was the organisation of the export of certified Demeter Biodynamic fruit to Europe, resulting from Podolinsky's 3-year struggle with entrenched interests and government bureaucrats. Eventually, the federal Agriculture Minister, John Kerin, a great support to Alex, circumvented the bureaucrats.

In 2008, Lynton and Joan bought an extra 35-acre (14-hectare) orchard block nearby, which borders the township of Merrigum. A prime consideration in purchasing this block and converting it to Biodynamics was to protect the residents of Merrigum from the chemical sprays that regularly drifted over the town. Lynton retired a few years ago, handing management of the orchard over to son Joel, with assistance from another son, Alex.

Walking over the orchard is like walking on a mattress. Photos can't do justice to the rich, compost-like soil or the beautiful rubbly structure of the cultivated ground.

Apple varieties grown are mainly Granny Smiths, Golden Delicious, Pink Lady, Fuji, Raeburn and Sundowners, while pears include Packham, Williams, Winter Nelis, Buerre Bosc, Josephine and Red Sensation.

Soil Management

Greenwood's is a good illustration of the minimal inputs required on Biodynamic farms in many situations. The only fertiliser input until Biodynamic conversion was one shovelful of poultry manure annually per tree. After Biodynamic conversion, Farry made Biodynamic compost for 5 years and applied it to the orchard in 1965–70. No fertiliser inputs at all were used after that until the late 1970s, when Lynton applied some cow manure one year. About 2001, he applied 150kg of reactive rock phosphate per acre (375kg per hectare) and the same amount of lime. He repeated this in 2002 but used nothing at all between then and 2008. Joel now uses small applications of guano at 40kg per acre (100kg per hectare) in sections every 3 years.

Green manure crops are the main providers of nutrients (via humus) to the trees. Two green manure crops are grown annually. In autumn, oats, trefoil and clover come up – usually from last year's crop, which is allowed to develop seed – and grow through winter. Usually, the crop is slashed in spring, irrigated once and allowed to regrow and fully express itself, maturing its seed. This is not ideal from a green manuring point of view but means the orchard doesn't have to be re-sown each year. It is then cultivated in with a home-made heavy disc plough that cultivates 100mm deep across the rows and along the rows.

If enough water is available, the areas green manured are flood irrigated 4 weeks after cultivating. This gets the soil boiling over with activity, consuming the organic matter and giving the trees a big lift. Another green manure then self-initiates, consisting of summer grasses, wild millet and *Setaria*. This grows for 6–8 weeks to 350mm high on two or three irrigations and is cultivated in mid to late December (summer). This crop takes from the trees in spring but gives the trees an enormous boost after cultivating. Pears grow more in the last 4 weeks before harvest than at any other time; this second cultivation gives the trees a boost at just the right time.

The summer grasses germinate again – there is still enough seed in the ground – and are allowed to set seed after the fruit is off the trees in March (autumn). The native trefoil germinates with this second crop of summer grasses and, as they die down in the autumn, the trefoils burst through: by May there is a good covering. In autumn the green manures are not cultivated, but slashed.

Prepared 500 is normally sprayed within 4 weeks of cultivation in spring, after irrigation, and in autumn, after slashing. A reciprocating spray rig is used, which puts out big droplets, the tractor driving down every second row. Preparation 501 is applied in spring when there is excessive rain or during long periods of overcast, warm, humid weather. Preparation 501 is sprayed out with a home-made sprayer that feeds the 501 liquid into a stream of air that finely atomises it, shooting it up 20m high.

The Greenwoods have tried growing the trees in permanent pasture, and this method can work well too, though it requires plenty of irrigation water. The pasture is slashed every 30 days to keep it growing dynamically. The difficulty with this method is to maintain the clover balance. There is a tendency for the clovers to gradually give way to grasses such as paspalum that directly compete with the trees.

Drought/Climate Change

A severe drought extended from the late 1990s until around 2010. Average rainfall in the area is around 400mm, but in 2008 it was more like desert rainfall. The first year of heavy water restrictions was 2002, although for the previous 5 years the orchard was restricted to 100% of normal water entitlement. In 2008 the opening allocation was only 4%, later increased to 11%.

The severe restriction on water necessitated changes in management techniques. Irrigation water had to be severely rationed. Lynton had to cultivate and leave the ground fallow for as much as 2 months to conserve the moisture in the soil, much as a wheat farmer might cultivate and leave the ground fallow over summer to conserve moisture. The summer green manure had to be left out so that the trees could go longer between irrigations. This was obviously not ideal and the soil became a little finer and less structured as a result. Some areas required a little manure because of the limitations on green manuring.

Pests and Diseases

One upside of drought is that springs are relatively dry, reducing the risk of black spot. Normally a preventative programme is used for black spot: spraying lime sulphur during winter or at budswell. The Biodynamic winter tree spray is also used during winter (see Chapter 23), followed up with

Figure 26.2. Pears in flower, under pasture with clover flowering.

Figure 26.3. Pears in flower, fallowed.

wettable sulphur until flowering. Ecocarb, sodium silicate and copper may also be used. Seaweed is sprayed post-harvest to aid the breakdown of the leaves. This generally keeps black spot under control. As long as the orchard is kept clean in spring, any summer rain will not pose a threat, but be a big bonus in fruit development.

A wide range of methods have been used against codling moth over the years, including neem, garlic, pheromone confusion, *Trichogramma* wasps, Dipel, and lights with water traps. In the past, pears struck by codling would ripen prematurely on the tree, causing problems at harvest. Pears are picked and stored hard, ripening when removed from the cool store. Ripe pears squash in picking bags and in grading machines, causing a lot of extra work. Now, codling moth is less of a problem. Codling moth strike has declined noticeably over the years. A home block of pears that was not treated at all also showed a marked decline in codling. Lynton says that the environmental vitality developed with Biodynamics over 60 years has greatly strengthened the trees. When a codling moth attacks a fruit, the fruit now has enough stamina to heal itself, will not ripen and is marketable.

Figure 26.4.

Figure 26.5. Packham pears.

Figure 26.6. Red Sensation pears.

Similarly, if there is a bit of black spot, it now doesn't distort the fruit as it used to; the pear will keep growing under the spot.

Joel Greenwood's current approach to codling moths encompasses Cyd-X (*Cydia pomonella granulovirus)*, a virus disease that attacks active moths; pheromone mating disruption; entomopathic nematodes (*Steinernema feltiae*) for biological control of codling larvae while they are overwintering; and *Mastrus ridens* wasps, which parasitise codling moth larvae.

The orchard is home to a healthy population of natural predators. Over the last 10 years a real balance seems to have come about in the orchard with regard to pests. The Greenwoods have very few problems with the insects that bother conventional orchardists, such as apple moth, San Jose scale, two-spotted mite, mealy bug, thrip, dimple bug and looper. Pear and cherry slug can be a bother at times. Lynton used natural pyrethrum (without piperonyl butoxide), but used

it very discreetly to avoid damaging the natural predator population. Up to 20% damage to the leaves from pear and cherry slug is regarded as acceptable. Too much damage has repercussions for the next season's crop. Again the progressive strengthening of the environmental health of the orchard seems to have brought a degree of control over this pest. Joel's current approach is to spray with neem oil when it is needed.

In recent years, Queensland fruit fly has moved into the area, a serious challenge to any organic grower. Joel deals with them by hanging fruit fly traps in the trees. These contain a protein attractant and a fumigant. The fumigant is contained in the trap, never comes in contact with soil or plants and is organically allowed provided it is safely disposed of off-orchard after use.

Marketing

A variety of distributors are used, the largest being the BDMC, which sends Greenwood pears and apples to most Australian states. Apples are becoming harder to market as more and more conventional growers convert to organics, bringing a glut to the market.

Greenwood's fruit is generally of very high quality and is in great demand. The pears have a beautiful, unique Biodynamic flavour and are slow to ripen. (The fast ripening of Williams pears can be a problem for retailers.) I first tasted a Greenwood pear in 1971 and still remember my amazement at the beautiful flavour. It was the first time I had tasted Biodynamic fruit. I stayed briefly with Farry and Audrey in the early 1980s and remember Farry showing me his annual test. On the mantelpiece of the kitchen were two saucers, one with one of his own pears, and one with a neighbour's pear. They had been there for several weeks. Farry's pear was still in perfect condition, but his neighbour's was a pile of mush. Farry told me he did this every year to test the quality of his pears. The apples are always beautifully flavoured, too, and in the best years they absolutely explode with flavour in the mouth.

The Greenwoods make some of the fruit into certified Biodynamic juices: pear, apple and apple/pear. They also coordinate with other Biodynamic growers to produce apple/lemon and apple/beetroot juices. The juices have no additives and are pasteurised at the lowest possible temperature.

Lynton says that the orchard has a life and character of its own and has had a profound impact on many people over the years. People write or visit years later and comment on how much they learnt about Biodynamics from

visiting and working on the orchard. The Greenwood family have hosted the BDAAA's annual conferences[1] at their orchard nearly every year since 1987.

FOOTNOTE FOR CHAPTER 26
1. Attended by up to 300 farmers (Western Australian and Queensland farmers often can't attend). Food is Biodynamic and is provided by the farmers.

Chapter 27

Biodynamic Vineyard, Bourgogne, France

Marc and Pierrette Guillemot produce Biodynamic Chardonnay on 6 hectares near Macon in Bourgogne (Burgundy), 400km south-east of Paris. This part of the French countryside, whether farms, gardens or villages, is a beautiful picture of care and attention, and Marc and Pierrette's vineyard is no exception.

The vineyard is in five separate parcels of land within a few hundred metres of each other. This kind of thing is common in France owing to the splitting of land for inheritance, and the reluctance of farmers to swap their holdings and consolidate them in one block. Marc and Pierrette have five rows of a neighbour's vines in their vineyard, near the boundary with the neighbour, but he won't swap!

The soils in this area are not suited to red wine grapes, only to white, and Marc and Pierrette grow only Chardonnay grapes. The vine rows are 1.5m apart, and the vines (30–80 years old) are spaced 1m apart in the rows.

The Guillemots converted the vineyard to organic methods in 1988 and commenced with Biodynamics in 1992. For the first few years they sought advice from German Biodynamic advisers and used preparations from the French national Biodynamic group. However, this was

Figure 27.1. Domaine Guillemot-Michel, Macon, Bourgogne.

not really satisfactory. The advisers said it was not possible to have only vines; if they wanted to be Biodynamic they needed to have animals, vegetables and fruit trees as well. And most of the advisers couldn't offer much practical advice on the cultivation of vines.

In 1996 Marc and Pierrette met Alex Podolinsky, who was advising in the Loire Valley.

Alex said, 'Yes, it's possible to have just vines. You have only a small property and no room for animals or anything else. You work with what you have.' Since then they have followed Alex's advice, with exceptional results. In the past, the regional Biodynamic group made preparations according to Maria Thun's advice. Now they work with the Australian Demeter method. A large group of farmers in France are using this method. The main adviser and preparation-maker was Pierre Masson, who worked with Alex for years and has helped many French farmers convert to Biodynamics. His son, Vincent Masson, now carries on the work.

Biodynamic Preparations

Before meeting Alex, Marc used only straight 500. As he uses no fertiliser at all on the vineyard, not even compost, the influence of the compost preparations was missing. Since they started using prepared 500 (which incorporates the compost preparations), Marc and Pierrette's soils have improved dramatically. Other changes in management included the establishment and management of grasses and legumes in the vineyard; previously, the ground was clean-cultivated. Two 3-hectare Italian-made (Australian design) stirring machines are used for stirring 500 and 501. While Marc sprays one batch of 500 or 501 (with an Italian copper knapsack), the next batch is being stirred. Marc has fitted his knapsack with a 500 nozzle on the back, which sprays large droplets behind him as he walks. He also made a two-pronged 501 extension to go on the back of the knapsack, which sprays fine droplets up in the air. Prepared 500 is applied to the whole property in

spring and autumn. Preparation 501 is used as required in spring, often two or three times, and once or twice in the week or two before harvest. It brings more sugar into the grapes and dries them a little. It is used twice during autumn if there is a lot of rain before harvest. It is not needed in summer; the east of France has hot, dry summers and very cold winters (–12°C to –15°C is common).

Marc and Pierrette's vines being so close to their conventional neighbour's – including five rows of the neighbour's vines in their vineyard – it is easy to make comparisons. As Marc prunes the vines to open them up before harvest, it was hard, when I visited, to see the beautiful structure of spring. In spring, his vines grow up towards the sun with lively structure, whereas his neighbour's hang down towards the ground. However, I could still see a difference in the plant structure and leaf form. The Biodynamic vines had a more upright, harmonious appearance. The grape bunch stalks on the Biodynamic vines were starting to mature and turn brown. The conventional stalks never mature properly, but stay green.

The soil difference was striking. The Biodynamic soil was browner, well structured and alive looking. The conventional was paler and hard and tight looking, with mosses growing on the surface (a sign of lack of air in the soil).

Vineyard Management

Marc uses a large range of implements in the vineyard, many of which he has made or modified for his particular needs over the years. The Guillemots were the first in the area to grow

Figure 27.2. Marc Guillemot.

organically. When they first started, everybody else used chemical weedkillers and it was hard to find suitable equipment for cultivating and mowing vineyards. Alex regarded Marc as the most creative equipment-maker in Europe. When Alex described a 500 nozzle to Marc, he went into the workshop and made one in 5 minutes. All the equipment is cleaned after use and returned to the shed.

Marc grows mixed pasture in between the rows. He aims for 50% legumes and 50% non-leguminous grasses. He finds that rolling and flattening the grass periodically is better than cutting, though some years he will cut it once. Every 3 or 4 years the non-legumes start to dominate, and Marc then ploughs the pasture in, 5–10cm deep, with a reciprocating spader mounted on either a small tractor or a larger

Figure 27.3. Brown, structured soil.

vineyard tractor. The larger tractor straddles the vines and two rows are plowed at once. Marc has established by trial and error the best ground speed to operate the spaders. Too slow results in too fine a working and consequent loss of structure, whereas too fast produces too rough a surface.

Figure 27.4. Copper knapsack set up for spraying 501.

Figure 27.5. Reciprocating spader mounted on a small tractor. Photo: Marc Guillemot.

The pasture between the rows is managed according to the season. In a wet summer the grass strip is left wider to use more water, whereas in a dry summer it is kept narrower to use less water. The under-vine area is treated differently: a very good leguminous grass (French – mouron) grows there in spring and autumn and dies off in summer; a different species grows through summer. Marc cuts the under-vine grasses five or six times a year with automatically retracting blades that cut the grasses 2cm under the surface. The blades retract when an electric eye sees an obstacle. Re-seeding of the under-vine and mid-row areas is not necessary.

The vines are pruned in winter. The prunings are put straight into a portable fire from which the ash falls upon the ground. The fire also keeps Marc and Pierrette warm as they work! Thus the minerals in the cuttings are kept in the vineyard, but any disease spores are destroyed. After pruning, the Biodynamic fruit tree spray (cow manure, kaolin clay, microtomic sulphur and sodium silicate) is sprayed on the vines with an innovative machine designed and built by Marc. The only other materials used in the vineyard are sulphur for powdery mildew and copper for downy mildew.

Harvest

A few weeks before harvest, the vines are pruned to allow more air in, to dry the grapes and to make picking easier.

Figure 27.6. Vineyard tractor used for spraying (sulphur and copper). A similar tractor is used for ripping and ploughing two mid-rows at once.

Figure 27.8. Pruning.
Photo: Marc and Pierrette Guillemot.

Figure 27.9. Marc Guillemot indicates the width of mid-row pasture in a dry year.

Figure 27.7. Mid-row pasture, under-vine mouron.

Figure 27.10. Marc Guillemot indicates the width of mid-row pasture in a wet year.

Figure 27.11. Grapes ready to pick.

Figure 27.12. Pneumatic press.

I visited on 18 September, just before harvest. Marc and Pierrette had just made the decision to start harvesting on 20 September, earlier than they had expected. Their Biodynamic vines clearly tell them when the grapes are ready: the stalks at the base of the bunches start to turn brown. This is a very important indicator of maturity for Marc. The grapes we tasted were beautifully sweet, like sultana grapes.

The grapes are harvested by hand with scissors, picked into 25-kg plastic boxes. The boxes are stacked on pallets, transported to the press and tipped into the press by hand. The harvest takes about 8 days, involving 14 pickers, four workers in the cellar loading and cleaning the press (it is cleaned with water and dried each time it is emptied), two workers to transport the grapes, and Marc. For lunch, the workers eat meals prepared by a local restaurant. The grapes are crushed in a Swiss-made pneumatic press in which air pressure pushes a fabric on to the grapes.

Winemaking

After crushing, the grape juice goes to a stainless steel tank and is left overnight to settle and clear. Nothing at all is added. Next morning, the juice is put into ceramic-lined concrete tanks. No sugar or yeast is added, only 1–3g of sulphur (depending on the year) per 100 litres of juice. The tanks are filled completely to exclude air. Any juice left over is put in a stainless steel tank with an adjustable lid that presses on to the juice and seals air out. The concrete tanks maintain a good temperature (20–22°C) in the fermenting wine, but the temperature can be adjusted if necessary by running warm or cold water through coiled pipes in the tanks.

The grapes from the different blocks and different varieties of Chardonnay are fermented separately as much as possible. Marc and Pierrette can then better evaluate the quality of each batch in relation to variety, season and soil

Figure 27.13. Fermentation tanks.

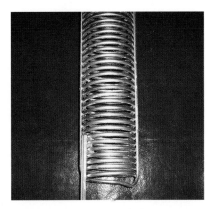

Figure 27.14. Coil inside fermentation tank for temperature control.

development. In the end, though, all the batches are blended before bottling.

Marc and Pierrette don't use wooden barrels because they and their customers don't like the resulting oak flavour. Their method produces a purer flavour, and growing the grapes Biodynamically brings a lot of extra flavour and aroma that they want to keep in a pure state in the finished wine. The wine is bottled 8–10 months after picking and stored in the cellar for 6–12 months before sale.

The wine is Demeter certified, and 90% is exported, mainly to the US, UK and Canada. Most of the export customers prefer the young Chardonnay, since they tend to drink it without food, as an aperitif. In this case it is good to have a beautiful young wine with plenty of fresh aroma. If you want a Chardonnay to drink with food, as preferred by most French people, it is better to have an older wine with more body and a stronger flavour. Marc keeps wine in the cellar from every vintage, going back 20 years. This is about the limit for Chardonnay, after which it starts to lose quality.

As I sped back to Paris (300 kmph) on the train, I was left with pleasant memories of meeting Marc and Pierrette and seeing their beautiful, vibrant Biodynamic vineyard domain.

Figure 27.15. Pierrette, John and Marc. Photo: Liuda Macerniene.

Chapter 28

Biodynamic Market Gardening, Latina, Italy

Agrilatina lies near the sea between Latina and Sabaudia, 60km south-east of Rome. This area was once a vast swamp. The draining of the swamp had already begun before Goethe's time. He observed this process and was inspired by it to write *Faust, Part 2*. The draining of the swamp was completed in the 1930s. Many eucalyptus trees were planted to assist the draining of the soil; this gives the area an Australian feel.

Agrilatina was founded in 1985 by Pasquale Falzarano, three of his brothers and two of his cousins. In 2011 the initial company split into two companies. Pasquale and his cousin Giovanni continued together with the Agrilatina company. In 2020 Giovanni died and Agrilatina is currently run by Pasquale and his wife Lucia, assisted by their sons Simone and Riccardo.

Nowadays Agrilatina's market garden and fruit-growing areas total about 80 hectares – 40 hectares owned and 40 hectares leased. It is a large vegetable and fruit-growing farm for this region. Normally, a farmer can make a living from 7–8 hectares. Agrilatina also manages another property of 78 hectares to breed cows, chickens and bees and to cultivate many kinds

Figure 28.1. Pasquale Falzarano, Agrilatina, Italy.

of heritage fruits as well as ancient cereals for farm bread production.

There are more than 40 workers and collaborators (some part-time) who have been trained, on courses held on the farm – in

Biodynamic farming, the structure of the soil, the objectives of the company and the broader social aspects of the farm's work, beyond the economic.

When possible, Pasquale is also involved in the spreading of Biodynamics and a sense of commitment to the wellbeing of and respect for people, environment and earth, by means of courses, seminars and conferences held in Italy and abroad, on farms and also in universities.

Furthermore, Agrilatina has a collaboration with a small village in the forest of South Sudan, where local people have learned how to cultivate the forest soil and to produce enough food as well as some products to cure their diseases. The people also have a good school and are learning some trades and many ways to work cooperatively.

Agrilatina was founded with the aim of producing large quantities of healthy food for people, because the quality of conventional food is very poor, and to address concerns about the earth's health. Another aim was to be of a size that could give it enough strength on the market to be able to deal directly with importers and some supermarkets.

The climate in this part of Italy is difficult. Although there are warm summers and mild winters, the proximity to the sea and to a coastal mountain (Mt Circeo, 541m) produces many severe storms with high winds, heavy rain, and hail. As Agrilatina wanted to grow many delicate vegetables, glasshouses were essential, not to force growth with warmth, but rather for the shelter from storm damage that they could provide. Agrilatina has 14 hectares under glasshouses, with the balance in outdoor market garden and fruit production. Fruit grown includes 18 hectares

of kiwi fruit and 3 hectares of many other fruits, mainly old varieties.

Most vegetables, especially the hardier ones, are grown outside; the more delicate ones are grown under glass. Vegetables grown include a wide range of lettuces, cabbages and chicories, corn salad (Italian – *valeriana*, German – *feldsalat*), carrots, radishes, Italian parsley, celery, rocket, chards, spinach, kohlrabi, tomatoes, leeks, spring onions, potatoes, watermelon, cucumbers, courgettes, squash, turmeric, ginger, melons, sweet potatoes and others.

The products are sold to distributors and some supermarkets in Italy and abroad. They are also made available to consumers in the farm shop (zero miles) to allow nearby people to enjoy very good products at a reduced price. In this way, healthy food can be truly accessible to everyone and not just to the wealthiest.

The first 2 years of Agrilatina history were transitional. The farming techniques were a blend of conventional and organic. Then, for 6 years, Agrilatina was organically farmed. These 6 years were very difficult, since the organic method did not adequately solve all the problems that arose and the organic market was not very developed at that time. The danger of losing the farm was serious. After 6 years of organic production, the organic matter content of the farm's sandy soil was 0.6%, exactly the same as when the farmers started converting to organic.

In 1993, Alex Podolinsky was advising in the area following a Biodynamic seminar in Tuscany. Camilla Conforti, who organised the seminar, was driving Alex past Agrilatina and asked if he would like to see an organic farm

Figure 28.2. Pasquale showing well-structured soil. Photo: Agrilatina.

Figure 28.3. Pasquale showing how the operator can see the crop at their feet, allowing precision weeding.

that might become Biodynamic in the future. Alex said he was very tired and would only look in for a minute. He and Pasquale talked, they looked at the soil and the farm, and Alex asked many questions. Alex said that most of what they were doing was wrong, which wasn't easy for Pasquale, but he recognised that Alex had a professional quality about him. Alex ended up staying all day.

After Alex left, Pasquale and his cousin Giovanni decided to go, almost immediately, to Australia. They weren't sure whether Alex was just a good actor or what he had been saying was real. For farmers, it is important to see with their own eyes, not just listen to someone talk. Alex took them to many Biodynamic farms in Victoria and they quickly understood that Australian Demeter Biodynamics was a very professional, successful way to farm. Giovanni carried a large package of Alex's 500 on to the plane home as his hand luggage.

Figure 28.4. Detail of cutting knives and crop-protecting discs on Fobro machine.

On their return to Italy, Pasquale, Giovanni and their partners started farming Biodynamically straight away. The whole farm operation was changed, with many phone calls and faxes to Alex for advice. They started spraying Alex's prepared 500 in autumn 1993, changed their cultivation methods and planted mixed green manures.

Figure 28.5. Soil preparation. Photo: Agrilatina.

Figure 28.6. Soil preparation. Photo: Agrilatina.

Figure 28.7. Seed sowing. Photo: Agrilatina.

They imported an Australian stirring machine. This was copied and is now manufactured in Italy. Alex trained some Italian preparation-makers for the purpose of supplying professional Biodynamic preparations to Italian farmers.

The Agrilatina partners decided to purchase, with considerable economic effort and a large debt, a 78-hectare property near Agrilatina, on which they have raised the beautiful, white Marchigiana beef cattle. Manure is collected, composted with straw and taken to Agrilatina market and fruit garden.[1]

Within a few years Agrilatina's soil had improved dramatically, as had the quality, health and flavour of the vegetables. The partners noticed this and so did their customers. This has been very gratifying for Pasquale and his partners. They saw the importance of high-quality, nutritionally rich food to people's health and regard the production of such healthy food as one of their main purposes.

After 6 years of Biodynamics, the organic matter content of the sandy soil had increased to around 4%, a very high figure for a sandy soil. Sandy soils contain high levels of oxygen, which tends to break down the organic matter quickly. This great increase in organic matter content is especially significant given the frequent soil tillage required for vegetable growing. It is much easier to build organic matter levels with permanent pasture than with market gardening. The formerly yellow sand is now a rich, warm brown colour. The high humus content of the soil has resulted in a much reduced need for summer irrigation of crops. Irrigation water is from underground; this region has plentiful water because it was once a swamp.

Fertility

Fertility is maintained and progressively developed with prepared 500 sprays three to four times a year (for each new crop and green manure), green manuring, Biodynamic compost and very careful cultivation.

When Alex first visited, Agrilatina was growing only mustard as a green manure, and only one crop per year. Alex suggested that at least 20 species should be used and that two green manure crops be grown per year. Agrilatina now grows summer green manures with up to 90 species and winter green manures with up to 30.

The composition of the mixture always changes and depends on many factors: the purpose of the green manure, the availability of seeds, their prices, the type of soil, the time of year, the climate, and how long you can have the green manure in the field. The exact composition is not as important as the knowledge of how the mixture works as a unique complex. In this creative painting it is very important to let many families work together, giving most room to the Leguminosae and then to the Graminaceae.

Alex points to the limitations of chemistry. For instance, the phosphorus content of a plant is determined by chemical analysis that assumes phosphorus quality to be the same whatever the plant. Alex suggests that this is akin to saying that the same note played by all instruments in an orchestra is qualitatively the same, whether played by a violin, clarinet or trumpet. Each different plant species contributes a different quality to the soil as it grows and as it decays. How boring would it be to listen to an orchestra consisting entirely of flutes? The wider the mix, the more enriched and enlivened will the soil be.[2]

Alex advised the farmers at Agrilatina, 'after cultivation, to sow down the green manure, to irrigate no deeper than 10cm and to walk out prepared 500. This brings the plants up and is followed by a second 10cm irrigation. The plants will come up to 20cm high. This is followed by one very deep watering, easy to achieve in a sandy soil, which in turn dries off from the top downwards. The roots, stimulated by prepared 500, follow the water deeper and deeper and the 500 activity therewith acts and structures deep into the subsoil. After four weeks the soil can thus be penetrated to 1m.'[3]

Glasshouse green manures are generally grown through the summer, because vegetable crops are grown in the glasshouse from autumn to spring, whereas, outside, green manures can be grown in summer or winter. If possible, two green manures are grown in succession before a vegetable crop, but sometimes only one is possible. When ready, green manures are chopped up finely with a mulcher before being worked (twice) into the soil with acutely forward-angled rippers followed by star wheels (the precursor of Australia's Rehabilitator plough – see p. 118).

The only fertiliser input used at Agrilatina is the cow manure and Biodynamic compost from the cattle property. The raw manure is sometimes used on green manure crops. No inputs are used on the cattle land, and, although manure and cattle leave the property, the soils and fertility levels are improving year by year. Prepared 500 is sprayed by knapsack on all newly sown or transplanted vegetable crops and green manure crops and after the green manures are worked into the soil.

Cultivation

In vegetable growing, the soil needs to be worked often, particularly to prepare seed beds for the many small seeds (e.g. carrots, rocket, corn salad, radish) that are to be sown. It is vitally important to retain soil structure, and implements must be chosen and used with extreme care. Most agricultural implements are designed by engineers who don't really understand soil and what is needed for the life in the soil, so Agrilatina workers often have to modify implements to better suit the needs of their soil. Most Agrilatina tractors and implements are set up to work 1.8-m bed widths.

Most of the cultivation is done with tined implements that move the soil gently without turning it and only work 15–25cm deep. Once a year the soil is deep ripped. For seed bed preparation, star wheels followed by a perforated roller gently prepare a fine surface for sowing, only working the top few centimetres and leaving the rubbly structure intact.

Agrilatina grows many of its own plants and produces some of the seeds it needs (e.g.

leeks, melons, rocket). Other certified seeds are purchased from organic or Biodynamic seed producers in various parts of Europe. Sowing and transplanting are carried out as much as possible according to the Biodynamic sowing calendar. Spring onions are sown or transplanted in both leaf and root constellations. Many vegetables are transplanted because direct-sown seeds sometimes germinate unevenly, leaving gaps, and weed control is also easier with transplants.

Weeds, Pests and Diseases

False seed beds are used for initial weed control. They are irrigated to bring up the first crop of weeds, which are then destroyed by cultivation. If there is time, there is a second irrigation and the second crop of weeds is also destroyed. If time is limited, the weeds can be flame weeded with an Italian-made gas weeder.

Once the crop is growing, a small, light, Swiss- or Italian-made tractor (20–40 h.p.) is used to weed between the rows. It has V-shaped slicing blades mounted between the front and back axles, in front of, and in full view of, the

Figure 28.9. Overnight holding yard for cattle.

Figure 28.10. Checking compost. Photo: Agrilatina.

operator. This enables very fine adjustments to the line of attack and the height of the blades. Discs either side of the blades protect the vegetables. When implements are mounted behind a tractor, the operator has to continually look behind, which is awkward and any small change in direction of the tractor can cause a big change in the position of the blades, resulting in crop damage. Weeding in the rows is done with small, sharp hand-held hoes and by hand close to the plants.

Pests and diseases are less of a problem than when the partners were farming organically, but are still present to some extent. Even the healthiest people sometimes get the flu. Agrilatina uses a range of allowable teas, oils and minerals (e.g. silica) to deal with problems that arise. Early sowings of daikon radish can be damaged by flies that lay eggs in the roots. The maggots then eat holes in the roots. These sowings are either covered with insect-proof netting until the weather cools off (when the flies disappear) or surrounded by 2-m insect-proof netting fences; the flies never venture very high. Later sowings, in cooler weather, are not troubled. Glasshouse cultivation brings the danger of fungus if there is too much humidity. The windows are left open all the time. The roof windows are open all day and at night are open enough to let air circulate but not enough for rain to get in.

Light

Preparation 501 (the 'light' spray) is used often at Agrilatina, particularly in winter when cloudy and rainy weather conditions prevail. In the glasshouses the light intensity is a little

Figure 28.11. Photo: Agrilatina.

reduced, and 501 brings more light to the plants. Summers are very warm and sunny, so 501 is not used so often then.

Beef Property

In 1997 Agrilatina bought the nearby 78-hectare grazing property to supply cow manure and Biodynamic compost to the main farm. Some vegetable crops and old varieties of cereals are also grown here. It was initially a dairy farm but is now used for beef cattle production and for cultivation. The more than 40 Marchigiana cow herd (and calves) are kept at night in partly roofed yards with deep straw bedding. Straw is regularly added to balance the manure and urine build-up and to soak up excess rainfall. Every 2 to 3 months the yards are cleaned out and Biodynamic compost heaps built with a manure spreader that has steel sheets attached to the sides to direct the material into a heap. No set ratio of materials is adhered to. Farmers have

Figure 28.12. Lettuce. Photo: Agrilatina.

Figure 28.13. Sweet potatoes. Photo: Agrilatina.

to look, to smell and to use their experience to determine the correct ratio.

All at Agrilatina are grateful to Alex Podolinsky for his ongoing help and advice. Pasquale observes, 'When you do something in a professional way it is very, very important

to work physically, to observe, to do with your hands, to understand, to try different ways, and arrive at the best way for your situation. These were inputs from Alex. We learned a lot from Alex, not only in farming, but also in general because he was a special man.

Figure 28.14. Harvesting rocket. Photo: Agrilatina.

Figure 28.15. Carrots. Photo: Agrilatina.

He was professional Biodynamic, but he had a very deep point of view for many things. That is very important for your life, for the planet's life. When we met him we noticed this, and we noticed it more and more over time. Now we are grateful to him for what he did. He gave us inputs to go on and it was very nice. He didn't tell you what to do, but you had to receive the inputs and then you were to act yourself. This is very different.'

Figure 28.16. Radishes. Photo: Agrilatina.

FOOTNOTES FOR CHAPTER 28
1. Compost-making is used at Agrilatina as an adjunct to green manuring and enables an extra crop to be grown each year. Many Biodynamic market gardens build humus with green manuring alone, compost not being essential.
2. Introductory text on Agrilatina 1 & 2 (DVD), BDAAA.
3. Podolinsky, *Biodynamics Agriculture of the Future*, p. 39.

Figure 28.17. Radishes. Photo: Agrilatina

Figure 28.18. Agrilatina bread. Photo: Agrilatina.

Pasquale Falzarano's personal reflections on the *Biodynamic Farming Handbook*

The book addresses, in an acute and appropriate way, the different aspects of the Biodynamic method as applied in Australia and in various parts of the world, especially following the impulses of Alex Podolinsky. The recognition and dedication addressed to him is also pleasing.

Alex Podolinsky was, for me and for Agrilatina, a turning point and a special relationship. With him we faced in depth many practical and subtle aspects of Biodynamic farming.

Over the years, thanks to the intense conversations, during his stays at my house and the many visits to Biodynamic farms that we made together, Alex has been a fundamental reference and stimulus in deepening the practice, thought,

experiences and evolutions of this method.

John Bradshaw, in this book, has managed very well to illustrate the experience and practices of Biodynamic farmers, making them accessible and useful for every type of reader. I also greatly appreciated the choice to expose the historical evolution and thought of the method, with the contribution that the various personages gave.

This, combined with a clear exposition of the preparations and useful elements, from the operations that can be carried out to the machines and tools that can be used, provides the reader with a very useful 'manual' which on one side allows everyone to connect with the spirit of Biodynamics and on the other side provides the suggestion of many useful elements to be able to put it into practice appropriately both in professional and hobbyist context.

Chapter 29

Biodynamic Grains and Legumes, Marnoo, Victoria, Australia

Steven and Tania Walter grow Biodynamic grains and legumes on just under 3,000 acres (1,200 hectares) at Marnoo, north of Stawell in western Victoria. Average rainfall is 418mm (16 inches). Steven's father bought the original part of the property in the late 1950s and Steven has farmed here since leaving school in 1981. His friend Terry Phillips (farming at nearby Callawadda) and he both hated using chemicals and when Terry came over to help with the shearing they often talked about how they could go organic.

One day, Steven read an article in *Acres Australia* about Biodynamics and Alex Podolinsky. He rang Alex, who said he would come to visit in a few days. Steven told Terry, who also rang Alex, and Alex visited both their farms. Terry started straight away, but Steven's father still had a lot of influence over the running of the Walters' farm and Steven had to wait a few years until he was able to take over. He started spraying 500 in 2000 and saw good results quite quickly. Within 12 months he could see the soil structure improving and the colour deepening in the lighter soils (it's not so obvious in the heavier soils). The colour and structure went deeper and deeper and all the soils have continued to improve

Figure 29.1. Tania and Steven Walter.

over the years. Steven and Tania also noticed that the wool became finer, dropping a micron or two fairly quickly. Their crops responded well. They had already started weaning them off chemicals and using natural fertilisers before they started Biodynamics, so the change was not too abrupt.

Steven and Tania's crops are comparable with their neighbours' in the dry years. Some neighbours farm conservatively with low inputs and have similar yields. The neighbours who use high fertiliser inputs get considerably higher yields in good seasons. However, Steven and Tania have very low costs and gain considerably higher prices for their crops, which are all sold as certified Biodynamic.

The year 2015 capped off a long drought with record low rainfall. In that year all the crops failed and soils were so dry they became powdery. To prevent long-term damage to the soil, the farm was destocked; all the sheep (Dohne/Merino crosses) were sold. The sheep had each been producing 6kg of 18–20-micron wool and good lambs, which were sold to organic and Biodynamic butchers in Melbourne and Sydney. The year 2016 saw a complete turnaround with a very good season – though too wet at times.

After the drought Steven and Tania decided to concentrate on growing grains and legumes, buying in sheep in a good season to eat off the excess grass and then selling them again. This brings much more flexibility in management. The sheep also help with weed management and recycling nutrients and waste from the

Figure 29.2. Mulching green manure.
Photo: Tania Walter.

Figure 29.3. Tractor with fluted discs.

Figure 29.5. Spray rig for 500 and 501.

Figure 29.4. Five-barrel stirring machine for 500 and 501.

Figure 29.6. Silos.

cropping system. With the sheep they were cropping one year in three (with pasture in between crops), but with no sheep they are cropping half the farm each year and putting the other half in green manure or leaving it fallow. The following year they crop the other half and green manure last year's cropped areas. The lighter soils need more time between crops to keep them in good condition.

When sheep *are* introduced, Steve and Tania prefer to get Demeter certified sheep from local Biodynamic farms where possible, because quarantine is not then required. If conventional sheep are purchased, they have to be quarantined for 3 weeks in a designated quarantine paddock, where they are fed Biodynamic feed and then put

out to pasture. The quarantine paddock is not used for any other purpose. The lambs that are bought from uncertified farms are sold on the conventional market.

Steven says, 'The benefit of not having the sheep on the farm all the time is that, before, I found that, rather than green manuring and stopping seed set for the cropping programme, I was looking for the feed for the sheep. We were compromising the cropping programme to look after the sheep, whereas, now, I do everything at the right time and it fine-tunes the cropping programme. It's especially better for weed control and getting a cleaner crop. The sheep don't do as good a job of knocking all the grass down as the mulcher does, so you end up having to mulch anyway to get all the weeds knocked over enough to stop them from setting seed. After harvest, the sheep pick up the residue from the crop. They eat all the summer weeds as well and, just by eating what's there, they're recycling some of their nutrients via their droppings.'[1]

Crops

Steven and Tania grow a wide range of crops. The largest area is in oats, which are processed into certified rolled oats, some of which are sold under their own brand: Burrum Biodynamics. Rye and spelt (an ancient grain related to wheat) are also grown. They are cleaned and bagged into 10- and 20-kg bags and sold to millers and bakers around Australia. Some is milled on-farm and sold direct to local people. Barley is also grown, bagged and sold for milling and to a distillery to be made into various products, including gin, vodka and whisky. Steven and Tania also grow a wide range of legumes: French green lentils, red lentils, faba beans and field peas.

Preparation for a crop begins a year in advance. Green manures are direct drilled in autumn – vetch, clover and medic along with a variety of other grasses that come up naturally. The green manure crop is topped during the season and finally mulched, worked in with fluted discs and fallowed to conserve moisture over summer. This programme helps restrict seed set, reducing the weed burden when the following grain or legume crop is growing. Steven used to cut hay to reduce weeds in the following crop but

felt it was taking too much organic matter off-farm, so he doesn't do that now.

Steven says, 'We have made changes with our harvest weed seed collection. Before, when we were harvesting the crop, the header would spread the chaff and the straw widely and the weed seed would go through the header and then get spread all over the paddock. We've now made some alterations to the header so that most of the weed seeds and some chaff get funnelled into a 150-mm-wide windrow and the straw still gets spread out over the paddock. This is called "chaff-lining". Many of the seeds will rot, but if weeds do grow they will be restricted to a very small section. Hopefully, this will get the weed management down to an acceptable level. As half the farm is in green manure and the other half is cropped – the two halves swapping over the next year – you have these very thin strips of weeds in the green manure following a crop, which will eventually be worked in.'

Grain and legume crops are sown in autumn or winter as soon as there is enough moisture. The soil is cultivated with tined implements – usually on the air seeder bar. Any fertiliser used is applied while sowing. Sometimes a little guano is used (20–32kg per acre [50–80kg per hectare]) to keep phosphate levels up. If there is a zinc deficiency, some zinc may be applied, but there hasn't been need for that recently. No fertilisers are added to the green manure crop. Prepared 500 is sprayed over the whole farm once a year when there is enough moisture in the soil. Steven would prefer to do this when there is plenty of soil warmth, in autumn, but there is usually not enough moisture until June or even July (winter).

Figure 29.7. Green manure mulched and disced.

Figure 29.8. Field peas.

A five-barrel stirring machine stirs enough 500 to spray 100 acres (40 hectares) per batch.

In 2021 Steven started experimenting with worm juice as a seed dressing. It is applied via a little plastic tube out of a 20-litre drum on to the auger on the air seeder. The seed then dries out fairly quickly before being sown. He's hoping that it gives the seed a bit of a head start out of the ground but at this stage it's very hard to quantify how well that is going.

Steven is experimenting with some multi-crops, including barley/peas and oats/rye, which can be harvested together and should result in increased yields of both crops, compared with single-variety crops. He hasn't yet had a great deal of success, owing to droughts and dry weather, but will try again in better seasons. He's going to try to put the different seeds in alternate rows and see if that works better. It will take a bit of work to set up the seeder to do so but he thinks it will be worth the effort.

While I was on the farm, we walked into the lentil crop. The soil was soft, like walking on a mattress. Conventional farmers attending a field day in 2016 compared Steven and Tania's soil with a conventional neighbour's and couldn't get over the difference.

Steven and Tania find that the Biodynamic crops are very robust and healthy. One year, they had very heavy rainfall in spring and the faba beans had a little chocolate spot on

Figure 29.9. Faba beans.

Figure 29.10. Faba beans.

the lower leaves. The chocolate spot had been showing for 3 weeks by mid spring, but hadn't progressed and the new growth was still green and healthy. Steven said that when they were farming conventionally, if they didn't spray chemicals as soon as chocolate spot showed up, the whole crop could go black in a few days. With Biodynamics, however, the plants can have some chocolate spot but continue to grow healthily without any treatment. Biodynamic preparation 501 would be ideal in these wet spring conditions, but the problem is that when it is needed the fields are far too wet to drive on to apply it, and when conditions are dry enough it is no longer needed.

An agronomist friend of Steven's visited in mid spring that year. He had been working with a farmer who was having a lot of trouble with fungal diseases in his chickpeas. They had already sprayed twice with fungicide (costing A\$20 per hectare each time) over several weeks, but the chickpeas were still full of disease and only 150mm high. The crop ended up dying. Steven went down to his chickpea crop and brought back a few plants. They were 300mm

high and completely free of disease. The agronomist was amazed and said, 'This just goes to show that you *don't* need chemicals to grow chickpeas.' Steven said that the health of the soil was a key factor and he was pretty sure the fungus the plants were getting was a sign of sick soil and that spraying fungicides would only make the soil sicker.

A native budworm (*Heliothis* grub) can do a little damage to the legume crops: it gets into the pods when they are drying off. Steven did use *Bacillus thuringiensis* (Dipel) in 2016 but hasn't needed to since, because damage has been minimal. Pea weevils can eat the seeds after harvest. To counter this, all the silos have fans that come on and off automatically depending on the temperature and humidity in the silos. These aerate the grain and cool it down, preventing pea weevils causing any harm. Carbon dioxide gas is used in the silos occasionally, but Steven finds that aeration works very effectively. Any grains that are processed are stored in a coolroom.

In 2016 some of the crops were flooded, including a spelt crop that had 150mm of water on it. However, only the worst-drained corner showed any sign of yellowing, and it recovered. Over the road, a conventional neighbour's barley crop, which had experienced comparable flooding, died.

There are very few organic farmers growing lentils in Australia, since they are a high-risk crop, susceptible to diseases and not competing well with weeds. They also require a higher pH than most legumes and need a free-draining soil.

Processing

Steven and Tania have developed a processing plant in a purpose-built shed to process their grains and legumes. Here they are able to produce and package whole and split red lentils, whole and split green lentils, yellow split peas, pearl barley, pearl spelt and soup mix. They also dehull spelt grain and process wheat. They have recently built a new storage shed to hold all the processed grain.

Figure 29.12. Burrum soup mix. Photo: Tania Walter.

It has a bigger coolroom and a packing area where Tania packs all the grains into various-sized bags, ready to go to market. There is also a small mill in there, and a new office. Processing machinery includes various seed cleaners and machinery for

Figure 29.11. Red lentils.

Figure 29.13. Steven and Tania's spelt that was flooded a few weeks before.

dehulling, splitting, pearling and bagging. The plant is fairly mobile, most of the grain elevators and bins being on wheels and easily moved as needed.

The processed grains and legumes are packaged in 1-tonne bulka bags, 20-kg bags and 750-g bags, with Steven and Tania's own Burrum Biodynamics label. They sell through the BDMC, which distributes to health food shops throughout Victoria and beyond. They also sell online directly: www. burrumbiodynamics.com.au. The BDMC and Mount Zero also repackage Burrum products and sell them under their own labels.

The processing facility has taken a lot of money and work to set up, including ongoing improvements to the production processes over time, but of course results in greater returns for the products.

Steven and Tania are very happy with their experience with Biodynamics. The steadily improving soils, healthy crops and value-adding give them much confidence for the future, together with the satisfaction of producing chemical-free, nutritious, life-giving food for consumers.

Figure 29.14. A neighbour's barley crop died off after similar flooding.

FOOTNOTE FOR CHAPTER 29
1.   Thanks are due to Biodynamic market gardener Mark Rathbone, who contributed to this chapter useful additional information about Steven and Tania's operation. For his audio version of my Biodynamic grain-growing articles go to https://www.saveoursoil.com.au/shop

Chapter 30

Biodynamic Free Range Eggs, Noggerup, Western Australia

Figure 30.1 Ed and Margaret Valentine, Noggerup, Western Australia.

Ed and Margaret Valentine began their Biodynamic farming journey in 1991 at Morowa, near Geraldton, in the northernmost part of the Western Australian wheat belt. They visited some Biodynamic farmers, read Alex Podolinsky's *Biodynamic Agriculture Introductory*

Lectures Volumes 1 and 2, attended a lecture by Bob McIntosh (Western Australian Biodynamic farmer and adviser) and began applying Biodynamics to their 3,750-acre (1,500-hectare) farm. They produced barley, noodle wheat, oats, cattle, sheep and wool and, together with a group of other Demeter certified farmers, supplied Biodynamic wheat to Japanese organic udon-noodle-makers.

As the Japanese economic crash, starting in 1990, gathered pace, this market dried up. Ed and Margaret began growing durum wheat, which was milled at the Biodynamic mill run by fellow Biodynamic farmers Dayle and Terri Lloyd. They found a pasta-maker in Perth, but couldn't source Biodynamic eggs and so started producing their own with 100 hens. After a very promising start, with delicious wholemeal pasta produced in Perth, a long drought set in. The Valentines decided to sell up, downsize and move south, where the rainfall was higher and more reliable. They bought a 360-acre (144-hectare) farm at Noggerup, 270km south of Perth, in a 700-mm rainfall area. The native forest extends from the back of their farm almost all the way to Perth.

A major feature of the farm is a huge dam, which not only looks beautiful but also allows irrigation of the poultry pasture over summer and of a citrus grove the Valentines planted. One of the two feeder streams of the dam is from Wandoo country, named for the *Eucalyptus wandoo* trees in the area, which is typically saline. The other stream is from a freshwater spring. The salty water from the Wandoo country stream sinks to the bottom, and Ed can lower the salinity by taking water off the bottom when autumn rains bring fresh water, and take the fresh water from the top for irrigation. He also has a device that suspends the salinity, enabling saline water to be safely used for irrigation.

The farm was in very poor condition when the Valentines arrived, with poor pastures dominated by Guildford grass. They very soon realised that their cattle had to go, since Guildford grass balls up in the cows' gut,

eventually killing them. The Valentines stocked sheep fairly heavily because sheep dig up the underground nuts of the Guildford grass in summer and eat them. Spraying 500 helped this process, since it made the soil more open and friable and made it easier for the sheep to dig up the nuts. This greatly reduced the Guildford grass over a few years. The pasture now consists of sub clover together with a mix of pasture grasses, native grasses and broadleaf weeds. Weeds are not of concern. Ed says, 'Weeds are not a problem in BD. Some years capeweed might be prolific. Next year the clover might be prolific. The following year it might be the grasses. Nature is continually replenishing itself by putting different plants there.'

Having made a start with hens at Morowa, free range egg production seemed to be a good

Figure 30.2. View over the farm and dam.

Figure 30.3. Free ranging hens.

Figure 30.4. Mobile sleeping and laying sheds.

Figure 30.5. Maremma dog on duty.

Figure 30.6. Older-style laying van: individual nests, from which eggs roll forward and are collected by lifting the covering boards.

Figure 30.7. Rollaway nests in laying shed.

Figure 30.8. Rollaway nest, AstroTurf floor, conveyer belt at the bottom. The wire at the back can be wound forward to expel hens at night so they don't sleep on the nest.

choice for their land. Ed, with help from his brother, began constructing mobile laying and sleeping sheds and fencing a 70-acre (28-hectare) field with electric netting. They felt that this would be large enough to move 3,000 layers around in, and it has proved to be ample.

As the pastures improved, Ed and Margaret reduced sheep numbers to 80 ewes and 70 wethers (plus lambs). They also transitioned from Merinos to Merino/Wiltshire crosses – which shed their wool – because it is hard to get shearers for small numbers of sheep. They had planned to phase out the sheep altogether and concentrate their efforts on the hens, but in 2015 the Australian government dealt a serious blow to WWOOF (Willing Workers on Organic Farms) and similar schemes.

WWOOF brings together organic farmers who need extra labour, and people who want to work on an organic farm for a time in return for board and keep. Many of the workers who have come to Ed and Margaret have been young people from overseas who wanted to qualify

Figure 30.9. Eggs emerge as the conveyer belt is wound by hand.

Figure 30.10. WWOOFers Alex (Belgium) and Chikako (Japan) grading eggs.

Figure 30.11. Sleeping van. The wire mesh floor stops foxes getting in.

Figure 30.12. Young birds.

for an extra 1-year working visa by working in country areas for 88 days. However, the government decreed that the 88 days must now be paid work, thus decimating the WWOOF and similar schemes. This measure has primarily affected organic farmers. There are still people wanting to WWOOF on organic farms, but far fewer, and the farmers and overseas visitors miss out on the wonderful cultural exchanges and long-term friendships that develop. The world so desperately needs the warm bonds of friendship, independent of culture, creed and religion, that build peace and harmony.

Genuine free range poultry production is labour intensive, and Ed and Margaret will have to seriously consider reducing their egg production and increasing their sheep numbers in the future.

Biodynamics

Ed and Margaret brought their four-barrel stirring machine with them from the property at Morowa. It is used for stirring the soil biology activator, prepared 500. Each barrel stirs 60 imperial gallons (273 litres) of water, with enough 500 for 20 acres (8 hectares), but they only need to use two barrels at a time here. Prepared 500 is sprayed over the whole property at least once a year, and sometimes twice. The best time is from autumn, after the first rains, until the rains stop in September or October (spring), provided the winter has not been too cold. The Valentines have seen the usual effects of 500: the soil gaining a lot more depth and structure, the grasses becoming more palatable and nutritious, and a great reduction in Guildford grass as the sheep became able to easily dig it up in the increasingly friable soil. Limited amounts of mineral inputs such as gypsum, dolomite, lime and a little guano assisted the conversion process in the early years.

There were some quite sick gum trees, infested with thrips, along a creek on the property when Ed and Margaret arrived. After a number of years of applying 500, the trees have become much healthier, whereas the neighbours' continued to decline.

Poultry

Ed and Margaret raise three batches of 800 day-old chicks (Hyline Browns) each year, in autumn and spring, avoiding the coldest and hottest parts of the year. The chicks are fed a combination of Biodynamic rolled oats, bran and pollard, rolled wheat, some brewer's yeast, sprouted barley, sometimes lupins, triticale and stillage.

Figure 30.13. Young hens, just starting to lay. Laying van to the left, jacked up level so the eggs roll properly. Sleeping van to the right

The stillage is a waste product from Biodynamic vodka-making in Perth. Biodynamic wheat is rolled and fermented with water and yeast. The liquid is pressed out and distilled into vodka, and the stillage is left. The feed regime is varied and refined according to observation of how the chicks are growing.

The chicks are kept in a fixed shed with access to a large yard that is covered with bird netting to protect them from crows and other predatory birds. They are given a wheelbarrow load of freshly cut grass a day when the grass in the yard is finished. When the chicks are about a month old, a mobile shed is parked close to their shed and they are allowed to explore it during the day. After a week, half the chicks are locked up in the mobile shed at night, while the others go back and roost in their original shed. After another week or so, all the chickens are sleeping in the mobile shed and can be moved to their first grass enclosure. They are kept here, close to the house where Ed and Margaret can

Figure 30.14. Young hens.

keep an eye on them until they are big enough to fight off crows. Here they start laying and become accustomed to laying in the mobile layer shed. Once they are in lay they are moved to the 70-acre (28-hectare) paddock with the older birds.

The hens are free to roam in the large paddock, wandering up to 500m from their mobile sheds, returning at night. Although there are some defections, most hens stay with their original home. The vans are hooked up in lines of three and moved to fresh areas when needed. This is done at night so that when the hens go out in the morning they reorientate themselves and can

find their home again. The layers are protected by a team of Italian Maremma guard dogs, who keep foxes and eagles at bay. The sheds are locked up every night as extra protection.

Ed and Margaret have planted many gum trees in the hen paddock, aiming to create a park-like atmosphere as well as provide shade and shelter for the hens. The trees were selected to be fast growing and have flowers suitable for honey, since Ed and Margaret keep a hive of bees. The trees look remarkably vibrant, undoubtedly because of the living soil in which they grow, and conventional farmers who visit remark that the farm has a very nice feeling.

Figure 30.15. Inside the sprouting van (not in use at present).

The layers are primarily fed whole wheat, stillage and shell grit, supplemented with sprouted barley over the hot, dry summer months, when the grass dries up. The sprouted barley replaces the fresh green grass and also makes protein more available. Supplying enough protein can be a problem for organic and Biodynamic free range egg producers.[1] Some producers milk cows (around ten per 1,000 hens) and make a simple yoghurt for the layers; others grow or buy in certified legumes such as field peas. Ed and Margaret find the sprouted barley useful and also feed the occasional sheep or kangaroo to the hens. Stillage also helps, producing more and larger eggs. Ed once saw an ingenious device that produced maggots as a protein source. Roadkill is placed in a cylinder accessible to flies. The flies lay eggs; maggots emerge and, wanting to get out and pupate, climb up a spiral channel on the inside of the cylinder. They reach the top and fall down a chute into a tub, where the hens can get them.

When Ed discussed the protein problem with Alex Podolinsky a few years ago, Alex suggested that spraying the pasture with 501 (the Biodynamic light-intensifying spray) would help by putting more nutrition in the grasses.

Feed bins filled with wheat are at the back of the sleeping vans. They are filled from outside but only accessible to the hens from inside the van in order to minimise losses to wild ducks. However, Ed scoops some wheat on to the ground each day because some hens don't like feeding inside.

Ed and Margaret bought a second-hand sprouting van some years ago. Each day over summer (and when a new batch of chicks are being raised), barley is spread in shallow trays at one end of the van. As each new row of trays is inserted the line of trays is pushed towards the other end of the van. A mister keeps the grain moist and as the trays approach the end they get more and more light. By the time they reach the end (4 days), they have substantial roots but just a small shoot (hens and pigs do better with a short shoot). The trays and van must be kept clean to prevent mould building up.

Eggs are collected twice a day, at 10am and 4pm Most of the laying sheds have rollaway set-ups whereby the laying floor slopes and the eggs roll down on to a conveyor belt. The egg collector simply winds the belt, bringing all the eggs out. They are transported to the grading room on a quad bike.

All the eggs are washed,[2] and then checked with a candler, before they are graded for size and packed. The eggs are packed in boxes of total minimum weight 540g (new layers), 600g, 700g and 800g.

Marketing quality Biodynamic free range eggs has never been a problem. People find Ed and Margaret without them having to promote the eggs. Margaret makes a delivery run to Perth once a week, distributing the eggs to IGA (Independent Grocers of Australia) supermarkets, independent grocers, organic shops and restaurants.

Owing to the disruption of the WWOOF scheme, Ed and Margaret may have to downsize their poultry operation as they get older, and increase their sheep flock. They also have a citrus grove, comprising oranges, mandarins and lemons, which are marketed as Biodynamic fruit. They have also bought a second-hand transportable commercial kitchen, complete with benches, ovens and a smoker. With the intense work of egg production, this has been on the back burner, but will no doubt be put to good use in the future.

Farmers like Ed and Margaret are so important in the life of any nation, revitalising nature with Biodynamic methods and providing food of the highest quality.

FOOTNOTES FOR CHAPTER 30
1. The (Australian) National Standard for Organic and Bio-Dynamic Produce allows the use of the amino acid methionine as a supplement, which complements the amino acids in grains to provide more balanced protein.
2. Some Biodynamic egg producers don't wash the eggs, but instead rub away any dirt with a good-quality kitchen scourer (e.g. Scotchbrite) and reject any with manure on them.

Chapter 31

Biodynamic Goat Cheese, Gidgegannup, Western Australia

Gabrielle Kervella bought a steep, rocky 250-acre (100-hectare) block 25km north-east of Perth in 1980. When she asked the Western Australian Department of Agriculture what she could do with it they replied that it was strictly a hobby farm and could never produce an income. She was, however, determined to do something productive.

The area still has much intact native bush, and its future is assured. A vast, privately owned nature reserve has been created, including a 16-km wildlife corridor. At each end a feral-animal-proof enclosure has been built, from which all the feral animals have been removed, and breeding programmes have been developed to reintroduce all the animals native to the area. If anyone in the area finds a creature they can't identify, they call the staff, who come out and identify it. If it's something that's missing, they take it away and breed it up. The whole reserve has been set aside for nature in perpetuity.

Gabrielle sold 200 acres (80 hectares) of her place in the mid-1990s and bought the 55-acre (22-hectare) block next door. Later, she sold off the remaining 50 acres of the original block to finance a new goat dairy and cheese factory on the new block. When she started organic

Figure 31.1. Alan Cockman.

farming she had to learn the hard way how to improve the poor, stony, compacted soils and it was sometimes a struggle to keep the goats in good health. In 1996 Gabrielle started applying Biodynamic methods to the farm with guidance from Bob McIntosh, the BDAAA's Western Australian adviser.

Figure 31.2. Gabrielle Kervella.
Photo: Alan Cockman

Soon afterwards, Gabrielle met Alan Cockman, an English cheesemaker who had spent 20 years in South Africa, including 8 years as a consultant to small, self-contained dairies. He was travelling in Australia and looking for work. Organic farming had always been an interest of his. Kervella sponsored his permanent migration to Australia as her cheesemaker. A relationship soon developed and they became partners.

Soil Development

Gabrielle and Alan realised early on that the impoverished soils would require some minerals to complement the biological development fostered by the 500. They added dolomite, lime and gypsum one year and rock phosphate the next year. When the truck came the first year they asked the driver to dump it all in a field for them to spread later. The driver asked if he would get bogged down. They said, 'No, it's really compacted.' During that year, they applied 500

three times and bounced and scratched over the stony fields with an S-tine cultivator to try to aerate the soil a little. They made Biodynamic compost with the goat manure and straw (the bedding material from the sleeping shed), leaves, green material harvested with a forage harvester, and whatever else they could find, and spread that over the fields. That year nothing really grew, but the next year, when the driver came with the rock phosphate, he drove on to the same field and nearly got bogged down. It was then that Gabrielle and Alan realised that they were starting to get somewhere with the soil structure.

They put in oats and barley for hay and got the hay contractor up to look. He said, 'The soil's dead, you'll have to put super [phosphate] on that, and you'll have weeds everywhere.' In late September, Alan phoned him and said he thought the hay was ready to cut. The contractor told him that nobody's hay was ready to cut yet, but that he would come and look. When he came he couldn't believe it. He said, 'What have you put on here?' They cut 600 bales from 16 acres (6 hectares). Next year they put in oats and field peas and again the hay contractor was amazed at the result.

Before starting Biodynamics, Gabrielle and Alan were feeding 3,000 bought-in bales of hay to the goats each year (they milk about 160 goats). With their own Biodynamic hay they have found that they only need 1,000 bales! The goats eat every bit of the hay, including the stalks, which were wasted before. The hay is always a sown crop (such as oats and triticale), since hay from permanent pasture is not feasible in this area. They cut early, before the seed gets to the milky stage, because they

Figure 31.3. The original impoverished orange soil has developed a more rubbly structure and darkened to a chocolate colour.

feel that the stalk is sweeter at that stage. They prefer a sickle mower or drum mower to a mower conditioner for best-quality cereal hay.

When Gabrielle and Alan started, the soil was a poor, tightly compacted, light orange mixture of gravel, granite dolerite and coffee rock which was lacking in organic matter and grew very little. Under Biodynamic management the soil developed a chocolate colour, became friable and each year produced 1,000 bales of hay from the 16 acres (6 hectares) of tillable soil. No further applications of fertiliser were needed after the initial applications of dolomite, lime, gypsum and rock phosphate. Preparation 500 is applied three times a year: at the first autumn rains, again in June (the first month of winter, when the soil is still warm) and a third time after the hay is harvested, in late October (spring).

The 500 is stirred in a 20-acre machine before spraying out. The spray rig covers an 18-m-wide swathe. Biodynamic compost is spread each year in autumn. Preparation 501 is not used here; Perth is one of the sunniest cities in the world, getting about 300 days of sun per year. Preparation 501 would bring excessive sun influence and dry up plants far too much.

Feeding

The goats are fed Gabrielle and Alan's own oaten (plus triticale, barley, field peas, etc.) hay together with a balanced ration supplied by Dayle and Terri Lloyd, Demeter Biodynamic farmers at Dumbleyung, Western Australia. This ration is made from Biodynamic bran, pollard, oat hulls and wheat (or other grain) together with some salt, gypsum and lime, and is pelletised. Some Biodynamic grain and lupins are also fed together with a mineral mix basically consisting of copper sulphate, sulphur and dolomite. No yeasts or hormones are fed to the goats. Some straw is bought from the Lloyds for the followers and kids.

Goats are browsers rather than grazers. Although they access pasture to some extent, this is limited, since worms can be a problem for goats on pasture. Gabrielle and Alan initially planted some tagasaste (tree lucerne), but it is goitregenic, causing serious thyroid problems for goats at certain times of the year. It is also a somewhat noxious weed and is undesirable near bushland.

By chance, they discovered that the daughter of a former Gidgegannup goat dairy farmer had done a thesis on *Acacia saligna*, a Western Australian native shrub, as goat feed.

Figure 31.4. Alan with coppiced Acacia saligna.

Figure 31.5. Milkers in shed.

Figure 31.6. Young does.

Figure 31.7. Mother and kid.

Figure 31.8.

They consulted her and began planting this shrub in non-arable fields. They have found it to be an excellent goat feed, if somewhat variable in nutritional content. They initially let the goats in to browse the shrubs, but found that they caused excessive damage, so now Alan harvests enough for each day after milking. The shrubs are effectively coppiced and easy to harvest. Gabrielle and Alan are progressively expanding the areas planted to *Acacia saligna*. In summer the goats mostly eat their own hay plus *Acacia saligna*.

Breeding, Management and Milk Production

The goats are trained to lactate for 700 days. Gabrielle and Alan breed from half the herd one year and the other half the next. Most births are of twins or triplets, with some quadruplets and quintuplets. Goats can look after twins and triplets adequately. The mums of quads and quins are sold off. The kids are traditionally taken away from the mothers at 4 days and bottle-fed, however Gabrielle and Alan leave them on the mums: the kids stay with their mums during the day and in the evening are put into a shed where they stay warm all night. The mums are milked in the morning, then reunited with the kids. The kids want to drink and this stimulates the mums to produce more milk. The mums wean the kids at about 6–8 weeks and the kids are then put in a separate pen. Gabrielle and Alan try to let them do it as naturally as possible.

The goats (mostly Saanen with some British Alpine) are milked in a fully enclosed shed, in a ten-a-side herringbone. The goats walk up steps to go to their milking station. During winter, milk production is only a fifth of what it is in spring, even though half the herd is still milking. However, they still manage to make good cheese. Over the year the goats average only 2 litres of milk per head per day. Gabrielle and Alan could artificially bump up milk production by feeding the goats rumen-stimulating yeasts but consider this to be unnatural and refuse to go down that path.

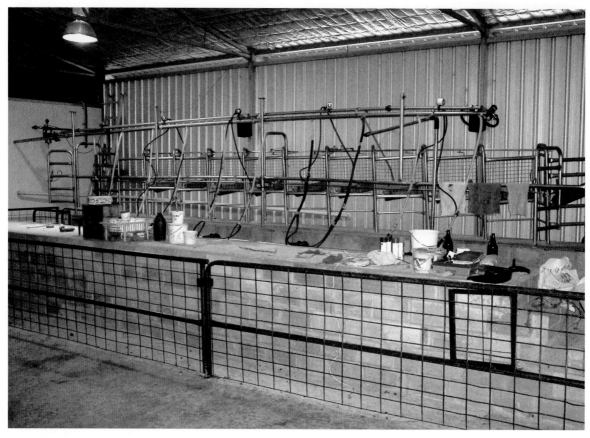

Figure 31.9. Dairy.

Although the milk yield is relatively low, the milk solids are extremely good. Gabrielle and Alan get very good yields with the cheese and also excellent quality and flavour.

The goats are not dehorned. In the past, when they *were* dehorned, the goats used to clash heads when fighting for hierarchy in the herd and come in with bleeding heads. Now, they clatter their horns around but never do any damage to each other with them. They use their horns to scratch their backs and all sorts of other things. Moreover, the kids grow stronger and better.

Cheese

Alan makes nine different cheeses, all based on the same curd, varying from a very fresh, very soft cheese (ready within a day), through semi-soft cheeses to the more mature cheeses. The three cheeses I tasted were of beautiful flavour, with none of the rank 'goaty' flavour associated with some goat cheeses. Kervella cheeses are distributed by different distributors in each state and are served to Qantas business class passengers. A few days after my visit I had lunch at a restaurant on the outskirts of Perth, where a most beautiful goat cheese was

Figure 31.10. Cheeses. Photo: Gabrielle Kervella and Alan Cockman.

served as part of a mixed platter. I had to ask, of course, and, yes, it *was* Kervella cheese!

Postscript 2024

Gabrielle and Alan have now retired from the farm and are currently making cow's cheese in New Zealand.

Chapter 32

Biodynamic Market Gardening and Cropping, Leipzig, Germany

Figure 32.1. Maria Bienert with Biodynamic glasshouse tomatoes.

Maria Bienert always had a vocation for farming. In 1989 she completed a 3-year organic and Biodynamic agricultural course in the Netherlands, which focused on vegetable growing, cropping and cattle breeding. She found the Biodynamic component interesting but too theoretical, mainly involving the reading of Rudolf Steiner's agricultural lectures.

After completing the course, she worked for a year on a New Zealand Biodynamic dairy farm. There she came across Alex Podolinsky's *Biodynamic Agriculture Introductory Lectures Volume 1*. It was the first time anyone had explained Biodynamics in a way that she could understand and that really made sense. The farmer she was working for told her that if she went to Australia she had to meet Alex. When she finally got to speak to him he invited her to the BDAAA annual conference at Greenwood's orchard in Victoria. She was deeply impressed to meet several hundred very committed and down-to-earth farmers who could all say from personal experience that Biodynamics worked.

On her return to Europe, Maria worked on a Biodynamic farm where she had to spray the Biodynamic preparations. The whole process was of very low quality: the preparations were dry and not transformed and not stored properly, and stirring was done in plastic vessels. She observed carefully over a period of time, but could never see any difference between areas that were sprayed and areas that weren't.

Having seen the results obtained by Alex's Biodynamic farmers in Australia, she knew that Biodynamics could work if done properly. After a period in charge of cropping and vegetable growing on an 800-hectare organic farm, she and her partner, Martin Haensel (who works off-farm but helps in his spare time), purchased a 27-hectare property near Leipzig. The property was in several separate parcels, but farmers in this part of Germany are prepared to swap land to consolidate in one block, and Maria and Martin were soon able to achieve that. Later they expanded the farm to 40 hectares.

Maria grows 3 hectares of vegetables; the rest of the land is in cereal crops (wheat and rye) or green manures. All produce is Demeter certified under the German Demeter standards.

The soils in Europe, in contrast to Australian soils, are still young and fresh. The last ice age was only 10,000 years ago, and it brought a lot of fresh material to the soils, such as crushed rock. The crushed rock has huge mineral potential that slowly becomes available. Maria finds that, with Biodynamics, her soils are steadily improving without needing any inputs – apart from a little manure from a neighbour's ten water buffaloes, which she composts for the greenhouse crops. Instead of external inputs, Maria uses green manures extensively between crops.

Biodynamic Preparations

Maria uses preparations from one or other of the preparation-makers trained by Alex (in Italy, France and Switzerland) and finds them very effective. She has sometimes been able to get preps from the BDAAA and finds them the best she has used. She has great respect for the effort, commitment, dedication, and painstaking accuracy applied to prep-making by Alex and his colleagues in Australia.

Maria studied preparation-making with Alex in Australia in 2006 and started making 500 in 2007 with a neighbouring farmer, Klaus Strueber, who has cattle and was also starting with Australian Demeter Biodynamics. Alex visited him and selected a suitable site for the 500 pit on his farm. Later, Maria and Klaus started making the compost preparations 502–506 on Maria's farm, since she has better, well-structured 'garden soil' on the market garden, better suited to making the solid compost preparations.

Maria recognises the importance of farmers working together cooperatively in prep-making. It is very difficult to make all the preps properly, each farmer only needs a very small amount, and the quality is better when large quantities are made in one place. You really need a decent amount to get the process going properly.

I asked about the problem of BSE (mad cow disease) in Europe and whether it was causing problems for prep-makers. Maria explained that there were no problems collecting and using horns or transporting finished preps around. The main problem was finding a transparent cow's mesentery for the dandelion preparation (506). Most cows in Germany develop fatty mesenteries; they get too little exercise because they spend a lot of time in barns and are fed too much grain and corn. The transparency of the mesentery is just one small detail among many that make the difference between high- and low-quality Biodynamic preparations.

Maria uses an Italian-made 4-hectare stirring machine (based on the Australian design) and also has a series of bowls of different sizes for hand stirring smaller quantities of 500 or 501. She uses prepared 500, 501, the compost preparations for compost heaps, and valerian as a warmth-bringing spray. Preparation 500 is normally used only once on each crop, but in hot summers, when plants are stagnating despite adequate watering, an extra spray will give them strength to get over the stagnation.

Maria uses 501 frequently. In this area there are many cloudy days and moist mornings. Pollution also reduces the available light. Maria sprays 501 on the vegetables every 10–14 days, and three or four times on the grain crops over winter. In autumn and spring it is hard to get the balance right: there are cold periods with little light intensity that demand 501, but, as 501 holds the plant back from over-lushness, there is a danger, in cold periods, of plants stagnating. Maria tends to spray valerian before the 501 at these times of year. This brings warmth to the plants and mitigates the effect of abrupt weather changes. Just as in Australia we have to be sure there is enough moisture in the plants and their environment before using 501, Maria has to make sure there is enough warmth influence before using 501 in spring and autumn.

Green Manures

Green manures, together with prepared 500 and careful cultivation, are the key to soil fertility on the farm. Maria has often observed that where a green manure grows well the following crop will do well, but if the green manure is not so good the following crop will struggle, even when compost is applied to compensate. This area, close to Leipzig, tends to be a bit dry and suffers

Figure 32.2. Green manure.

severe summer droughts every 3–4 years. This makes summer green manure crops grown in preparation for wheat or rye a bit tricky.

Maria prefers to grow a wide variety of green manure species, but is increasingly finding that the water-demanding clovers fail to grow successfully without irrigation. She finds that lucerne (alfalfa) is a much more reliable green manure. Its roots go very deep (to 14m) and can access ground water. It struggles in the first year but grows prolifically in the second, even through very dry periods, and turns poor sandy spots into high-yielding spots because the following crop can follow the deep root channels. Lucerne is a traditional fodder crop in this area of Germany, since it is very reliable under dry conditions. It doesn't do much in the top 20–30cm of soil, however. Green manures grown in the vegetable areas can be much more diverse, since they can be irrigated. A diverse green manure is the ideal; each species brings something different to the soil. Lucerne is generally omitted from irrigated green manures, since it tends to dominate.

Green manures are mowed with the tractor – except in the glasshouse, where a hand mower is used – before they are worked in. Lucerne is quite difficult to work in. Maria first breaks the soil open with an Agrowplow and then chops the lucerne with a rotary hoe. She goes slowly to cut as much as possible. If vegetables are to follow, she will sow oats after working the lucerne in, and ridge the soil (as for potatoes). The oats grow and suppress some of the regrowth, and after the winter frosts 95% of the lucerne is dead. Vegetables are planted in spring, and after the first crop a second vegetable crop can be grown. After the second crop, there is still time for another green manure before winter, this time including a greater diversity of species, which thrive under irrigation. Summer green manures in the glasshouse consist mostly of buckwheat and oats, since many species don't cope with the heat.

Implements

Maria has a wide range of implements at her disposal. She has tried many implements over the years and has found that most are fine provided they are used with discretion, at just the right time. Some of her implements are:

- The Agrowplow – used to break open the soil before further work is done, often with a rotary hoe. It is perfect for a sandier soil, but is capable of doing a lot of damage in Bienert's heavier soil. It creates big lumps if the soil is dry, but can safely be used when there is enough moisture.

- The rotary hoe – generally used with a quite high tractor gear and as low engine revs as possible without stalling the tractor. This results in the tractor moving fairly quickly over the ground, but the blades rotating at as slow a rate as possible. When the rotary hoe is used in this way, at 50–100mm depth, minimal damage is done to the soil structure. For taprooted weeds and lucerne it must go a bit slower over the ground to ensure enough roots are cut. It is used for working in green manures and incorporating green manure seeds that have been broadcast. Only one cultivation is possible with the rotary hoe. Any more will reduce structured soil to dust.

- Discs – only used when there are massive amounts of straw that need to be cut and would clog any other implement. There is up to 5 tonnes of straw per hectare after harvest. They don't work the straw in very deeply; instead they break and mix it shallowly. Discs are only used in summer, when the soil is really active.

- Mouldboard plow – only used in soil preparation for carrot crops. Maria goes very slowly with it. It creates a uniform soil with no mulch material on top. This allows the frost to penetrate deeply into the soil. Frost has the same effect on the soil as ripping: it cracks the soil. Maria finds that the carrots grow longer when the mouldboard plow is used.

Cropping

After green manuring, wheat and rye crops are sown in late September (autumn) and grow slowly through the winter. Winters here are cold, with frequent frosts, but not a lot of snow. Wheat can grow from 1°C upwards. Spring-sown grains would produce only half the yield, owing to the lack of moisture over the summer. The grains are harvested the following year in July or August. Harvesting is done by a neighbour who can do it quickly and efficiently – which is essential because there is often rain around harvest time. If compaction results, a green manure including faba beans, peas, vetches and oats will fix the problem. The grains are sold to a Demeter growers' market corporation in northern Germany, which sells them on to companies that make Demeter cereals and bread. After harvest, there is time for a short-duration green manure before the frosts start in November. If the season looks promising, peas and beans will be included in the mix, but if it looks as though it will be dry, a plant whose seed is cheap, like phacelia, will be used instead. If it fails, it will at least not prove too expensive.

Market Gardening

A wide variety of vegetable crops are grown, including beans, carrots, parsnips, diverse lettuces and salad greens, Brussels sprouts, kohlrabi, spinach, onions, leeks, Florence fennel, celeriac, tomatoes, capsicums, parsley, basil and cucumbers. Strawberries are also grown.

Some crops are grown for Maria's 160 vegetable box customers; a set mix of vegetables is provided to each customer each week. An organic farmer colleague picks up the boxes from Maria's farm each week and delivers them along with his own produce. This suits her well, since she wants to be farming, not delivering.

Other crops are grown for an organic/Biodynamic wholesaler, who also picks up from the farm.

Weeds in the vegetable crops are controlled by a variety of means, including duck's-foot knives mounted in front of the tractor driver. For some crops, a gas flame weeder is used to kill newly germinated weeds just before seedlings emerge.

Maria has no real pest problems except with leeks, which are troubled by small insects that lay their eggs on the leaves. These have become a problem since summers began to get warmer. Maria puts insect-proof material over the leeks because she hasn't yet found a better way to deal with them.

She has abundant ground water for irrigation, from an 8-m and a 14-m well. She uses sprinklers in the open fields and drippers and sprinklers in the glasshouse.

For carrots, Maria ensures that the soil is moist before sowing, and ridges the soil for more depth. The tractor-mounted seed sower places the seeds 2cm apart – they are not thinned – and a press wheel with a hard rubber ridge presses the soil firmly over the seed. Small tines then push a thin layer of loose soil over the pressed soil. This effectively mulches the carrot seed and prevents the drying out or crusting of the firmed surface. Good germination results, provided the weather is not too hot, too dry or too wet. Maria grows several local, open-pollinated carrot varieties for the wholesaler and tries to educate consumers that open-pollinated is best, even if there is more variation in size. Most carrots grown in Germany are now hybrid varieties.

Figure 32.3. Flame weeding. Photo: Maria Bienert.

Figure 32.4. Maria flattening the tops of raised rows, exposing moist soil for the seeder to sow carrots into. Photo: Maria Bienert.

Preparation 501 is sprayed in the morning as necessary and also 2–3 weeks before harvest, on a sunny afternoon. This supports maturing. The carrots are harvested by hand after drawing a blade through the soil at a depth of 30cm. They are stored in a coolroom at 1°C and 98% humidity.

Glasshouse

When I visited in early September, Maria was growing tomatoes, cucumbers, capsicums and basil in the glasshouse. She grows hybrid tomatoes, having found that yields of open-pollinated varieties were not satisfactory in the long term. From 600m² she produces 6 tonnes of tomatoes. The tomatoes are on a 4-year rotation to prevent build-up of diseases.

Figure 32.5. Carrot harvest. Photo: Maria Bienert.

Maria cultivated the soil once to relieve compaction when she first arrived, but has not done any cultivation, or taken a tractor into the glasshouse, since. Green manures are the main source of fertility, together with small amounts of Biodynamic compost made, with an old manure spreader, from the neighbour's water buffalo manure, straw and hay. Maria spreads Biodynamic compost 3–4 months before planting and sows the green manure. When ready, it is scythed or mown with a lawnmower and left on the surface. A little extra compost is then spread, just on the rows. As soon as there is no danger of frost in the glasshouse, the seedlings are planted. Prepared 500 is sprayed, followed by 501 a few days later. (Repeat applications of 501 are made during the

Figure 32.6. Glasshouse cucumbers during cooling conditions late in the season. Some mildew is appearing, but one cucumber per plant is still being picked every second day.

growing season, sometimes as often as every 2 weeks.) The plants are watered well for a week, then the soil, green manure and compost are covered with perforated plastic weedmat. Maria finds that the worms and other soil life take everything down into the soil, which now has a beautiful structure and 5% organic matter content (it was 2% initially). Although weedmat is not ideal, it does allow the soil to breathe adequately, stops weed growth and reduces the amount of water required. Irrigation is applied via dripper line laid under the weedmat. Wooden frames 300mm wide are laid on the pathways to prevent compaction from walking.

Maria finds that the flavour of the vegetables is highly dependent on the quality of the preceding green manure crop. The better the green manure, the better the flavour. The tomatoes I tasted were beautifully flavoured and the red capsicums were absolutely sensational.

With the weather starting to cool off, the glasshouse cucumbers were starting to develop mildew. Some artificial heat would prevent this, but Maria doesn't heat the glasshouse at all. She grows a variety of cucumber that tolerates a bit of mildew. Although the season was drawing to a close, the plants were still giving one cucumber each every second day.[1]

Seed Saving

Maria grows seeds for a German Demeter seed company, typically onions, lettuce, carrots, spinach and herbs. Onions are a challenge, because they take so long to produce seed, and are at risk of collapsing. They and other non-fruiting vegetables have been bred to remain in maturity as long as possible and not continue to seed. Maria sprays 500 when the plants are losing energy, and this helps them pick up the growth impulse again and carry through to produce seed.

Figure 32.7. Onion seed drying in the glasshouse.

FOOTNOTE FOR CHAPTER 32
1.   More detail on Maria's tomato and carrot production can be found in *Biodynamic Growing*, 16 (2011), pp. 23–26 and *Biodynamic Growing*, 20 (2013), pp. 8–11, respectively.

Chapter 33

Yatesbury House Farm, Wiltshire, United Kingdom

Text and pictures by Richard Gantlett[1]

Figure 33.1. Yatesbury House Farm.

While reading this, you can be happy in the knowledge that over 1,000 barrels of Demeter whisky are ageing on the Isle of Islay in Scotland as a result of two Biodynamic barley harvests on our farm. Last year's harvest of 2013 is being turned into malt as pen is put to paper, soon to add another 400 barrels to the store. Our 12-year-old son is looking forward to tasting the outcome. It will be legal for him to do so at the next tasting in 10 years time!

I first came across Alex Podolinsky while our farm was halfway through organic conversion in the late 1990s. Research into soil-enhancing and -enriching methods had led to the book *Secrets of the Soil.*[2] Alex featured in one chapter of this book and I was fortunate enough to spend a week later that summer following Alex from farm to farm while he toured France and opened the eyes of like-minded farmers, growers and viticulturalists. It was Alex's determination to show that the Biodynamic method really changed soils and plants that inspired me. The vineyard of Guy Bosard was where I first clearly saw Biodynamic soils and plants, completely different from the neighbours' non-organic

Figure 33.2. Richard Gantlett.

models. Since then, inspiration has come from many of our group of European Biodynamic trailblazers but particularly from visits from Frances Porter[3] and Alex.

We have a pedigree black Aberdeen Angus herd from which we sell half as breeding stock and half to a local hotel. The hotel is organised and motivated enough to take two whole animals every 4 weeks, a perfect arrangement for any farmer with a suckler herd. Angus, particularly the low draft ones, are good at forage conversion and even before organic farming caught on here our cattle were not to be fed cereals. It made no sense to feed cereals to our cows. Instead rotational grazing is practised, which benefits cow and soil.

Figure 33.3. Pasture in full growth.

Our new farming system, which was emerging in the mid-1990s, had to be sustainable for future generations, as I told a farm adviser who visited us at the time, not quite knowing what that would mean then. Now the word 'sustainable' seems rather overused. The suckler herd started at the same time as organic conversion in 1998; the last of our farm went into conversion in 2003.

In the 1990s, before our change of system, cultivators were used to till our silty clay loam soil and rotational ploughing to bury straw

Figure 33.4. Research plots at Yatesbury House Farm.

stubble and grass, though this ploughing all too often resulted in concrete blocks that were of little use in cereal growing. Weaned off the plough with a little encouragement from Alex and the Wenzs from Germany, we sold the plough, knowing that otherwise it would be tempting to use it. This was quite a radical approach to organic farming in England at the time. Both the water-holding capacity of the soil and the time window of appropriate soil conditions are much improved, particularly for spring sowings where these factors are critical. As a result of our positive outcomes, there is now a research project looking at organic

Figure 33.5. Crimson clover in pasture.

farming without ploughing on other UK farms. A modified Italian ripper/lifter or 'Pinocchio' works beautifully to lift the soil gently, relieving compaction when necessary.

Weeds are always a challenge on all soils and farms. Perennial weeds – which I was most concerned about before conversion – are now my least concern. Wild oats have become a problem in some fields, or an indicator of the improved soil conditions, particularly the openness our soils are developing. A combination of approaches now has the situation under control. Other weeds such as charlock are good indicators of work needing to be done or a rest needing to be taken.

Field beans and wheat are grown for a local organic dairy farmer, and milling wheat and spelt for local flour millers. Each fits into the rotation depending on the field and year. In very wet years we would almost prefer to give the whole farm to pasture. It is amazing how inversely different the English climate is to the Australian one: often, when one is in drought the other is suffering excessive rain.

Fourteen hundred acres (570 hectares) are farmed here at Yatesbury, with a further 200 acres (80 hectares) rented at the hotel mentioned above. Yatesbury has been farmed Biodynamically for several years starting 2004/5. The whole farm was fully Demeter certified for the 2009 harvest. Our present rotation is pasture for 3 years, winter spelt, winter wheat and bean mix, green manure, spring wheat, beans, and spring barley undersown with pasture.

Does it work? Have our soils improved? Do our plants look different? Is the quality of our produce better? Well, the easy answer is 'Yes'.

Figure 33.6. Wheat and bean mix.

Figure 33.7. Alex Podolinsky examining a bean crop.

Figure 33.8. Agroforestry project at Yatesbury House Farm.

Figure 33.9. Heritage wheat on a grey day.

Figure 33.10. Alex 'sampling' Richard's whisky in the barley field.

Figure 33.11. Yes, the sun does shine! Harvesting barley undersown with pasture.

Ask anyone who knows the farm and they will tell you it has been totally transformed. The quality of our beef is improving all the time, and our main customer is no pushover. Our Scottish whisky distiller, when asked about the quality of the distillate from our malting barley, said it was 'smooth and milder than from other samples and gave a higher yield'.

The more complicated answer would be to try to identify causes for the improvement. Biodynamics is a systems approach to farming, and therefore you cannot separate out one piece: the preparations, the lack of ploughing, the closed farm, the diverse pastures, the cows, mixed varieties, mixed species cropped together, reduced size of machinery – all the factors contribute positively, including the people. Take one away to analyse and you lose the point, lose the whole system, which is bound to be greater than the sum of the parts. For those mathematicians among you, it is John Nash's 'equilibrium', also known as 'governing dynamics', applied to farming.

The diverse pastures mentioned above are really important to soil improvement and stabilisation. We use a mix of 19 species, within 35 varieties, of legumes, grasses and herbs to feed the soil and cows. This was one of Alex's approaches[4] which we adopted early on. It made sense to colonise the whole soil profile with different plant roots; our soils vary markedly across fields. With our seed mix we have plant seeds available for all our soil types, and the cows have a mixed herbal sward to graze.

We also take this principle into our cereal crops where possible, using mixed varieties of wheat and also growing wheat and field beans

Figure 33.12. Bluebell wood.

together in the same field at the same time. According to doctoral research elsewhere, a 50% normal sowing of each seed results in a 60% normal harvest of each. Yatesbury House Farm results show less weed competition, less disease and better harvests than with separately grown wheat and beans. These differences have not been quantified, however.

Still on the theme of mixed species, we have been growing a mixed variety of spring wheat known as 'heritage wheat'. The seed was originally collected from the understorey of thatched roofs during re-thatching and was then multiplied up to field scale. The result in the field is a picture of varying wheat plants, tall and thin-strawed, a reasonable yield and nearly no weeds. A combination of growth habit and, probably, root interaction has a distinct effect.

Perhaps these characteristics will be important for future plant breeding?

The interaction of ideas and the sharing of knowledge are essential if we are to move organic and Biodynamic farming along. The European meeting of Podolinsky-inspired Biodynamic soil developers, which passionately shares experiences and progress, is a wonderful space for this. If we are to spread these methods much more widely, there is a desperate need for plant breeding research to bring forward the traits in plants, particularly annual plants, appropriate to sustainable, natural, agro-ecological and Biodynamic farming methods.

Yatesbury House Farm Revisited

A great deal has happened since the first article was published about the farm transformation in 2013. The Yatesbury House Farm website follows the evolution and developing philosophy. Recent projects include: PhD on farm research; carbon auditing; farm health and public goods assessments; agroforestry as part of the whole farm organism approach; expanded use of heritage cultivars; and many projects sharing techniques and principles of the Yatesbury Biodynamic method.

A 2019 carbon audit exceeded all expectations, showing that Yatesbury House Farm was sequestering (storing) 10 times more carbon than it was emitting. The carbon audit was precipitated by speculation about the impact of methane emissions from cows on climate change. Cattle are known to support soil fertility, insect diversity and encourage farmland birds, all of which were enhancing the health of the farm as a whole, now the carbon impact was a little clearer. The 280 head of cattle were indeed contributing 73% of the farm emissions, calculated as carbon dioxide equivalent (CO_2e) emissions; 14% of the emissions were from biodegrading crop residues in the form of nitrogen dioxide (NO_2), and 10% were from the use of diesel fuel for tractors and crop drying. Of course, Yatesbury House Farm has no emissions from chemical fertiliser and pesticide production or use. Woods, hedges, and field margins offset 35% of the farm emissions, soil organic matter results indicated a doubling of soil organic matter since organic conversion. Hence the substantial carbon sequestration on the farm. Although this was really exciting and important, the evidence was not irrefutable, since the research was not strictly rooted in the scientific method

of experimental replication and reproducibility. Fortunately, this was about to be addressed.

In 2014 an on-farm experiment had been set up to investigate the effects of returning all the crop residues back to the soil. Alex had espoused growing as much diversity as possible in leys and green manures: 'It doesn't matter what plants you sow. Just sow as many different species and varieties as possible.' So we have been using diverse leys since 2000.[4] Alex also advocated returning as much biomass to the soil as was available. In the Yatesbury district it is common for arable farmers to sell straw to livestock farmers for them to use as winter animal bedding. Other demands, including mushroom compost, provide a market for crop residues to be sold off-farm. Hence the experiment to investigate the benefits of returning crop residues to the soil. This experiment was the basis of my PhD at the University of Reading. The experiment ran for five years until autumn 2019; it took another 18 months to fully analyse and interpret the results. When the experiment was set up in 2014, changes in soils were seen as long-term processes and therefore it was assumed that 5 years would probably not be enough to register scientifically significant changes. However, analysis of the data confirmed the earlier carbon audit results. The experiment had two main conclusions: firstly, carbon was seen to be accumulating in the soil as soil organic matter, at rates two to three times faster than the difficult to achieve[5] global target of increasing soil carbon by 4 per 1,000 (0.4% per year); secondly, although a peak soil carbon level for the farming system was found, the peak could be raised substantially by extending the length of the ley phase in the rotation. This is valuable for both climate change mitigation and adaptation: storing carbon in the soil removes greenhouse gases from the atmosphere but also improves the functioning of the soil in terms of nutrient cycling, water storage, pH buffering, and ease of tillage.

It must be emphasised that it was never the goal to increase soil organic matter on the farm. Rather, the primary purpose was to improve soil function by increasing the life in the soil.

The development of trees within the farm system started with allowing cattle to graze in small copses next to permanent pastures. This was so successful in enabling cattle to shelter from sun and rain, to browse more diverse nutrition, and also to increase biodiversity under the trees in consequence of cattle activity, that the Yatesbury Forest Farm was envisaged. The Yatesbury Forest Farm approach seeks a unification of the whole farm, including woodland, hedges, field margins, pastures, and crops, by means of the cattle's herd instinct. When the diverse leys are growing quickly then rotational grazing is adopted to ensure the full regrowth of developing plants especially their roots. However, as growth slows, cattle are allowed access to greater and greater areas. In autumn, it is envisaged that the cattle will access the whole farm, only the newly planted crops being fenced off to protect them. A silvo-cereal cropping system has also been planted to explore the cropping of top fruit trees in rows and alleys with a cereal-diverse ley rotation.

Heritage cultivars of crops have been increasingly grown in order to preserve genetic diversity, explore their dynamics with wildlife,

and examine their value for human nutrition.

The term 'farm homeostasis' encompasses the beauty of this holistic approach in which the cattle, the diverse leys, and the thriving soil life, enables the self-organising of nature. In summary, the development of the Yatesbury Biodynamic farm organism came about from seeing inspiring methods elsewhere and from exercising the imagination to bring those ideas together at Yatesbury in one whole organism. Although many of the practical approaches were intuited, they have been validated through the scientific method.

Text and Images by Richard Gantlett PhD, Yatesbury House Farm, Wiltshire. Visiting Fellow in Agriculture at the University of Reading.

FOOTNOTES FOR CHAPTER 33
1. This chapter was written by Richard Gantlett for *Biodynamic Growing*, 21 (2013) and is included here as an example of Australian Demeter Biodynamic method developments in the United Kingdom.
2. P. Tompkins and C. Bird, *Secrets of the Soil*, Penguin, 1992.
3. Frances Porter, Certification Officer, Bio-dynamic Research Institute, Powelltown, Victoria (1992–2016).
4. Döring, T., John A. Baddeley, Robert Brown, Rosemary Collins, Oliver Crowley, Steve Cuttle, Sally Howlett, Hannah E. Jones, Heather McCalman, Mark Measures, Bruce D. Pearce, Helen Pearce, Stephen Roderick, Ron Stobart, Jonathan Storkey, Emma L. Tilston, Kairsty Topp, Louisa R. Winkler and a. M. S. Wolfe (2013). "Using legume-based mixtures to enhance the nitrogen use efficiency and economic viability of cropping systems."
5. Minasny, B., B. P. Malone, A. B. McBratney, D. A. Angers, D. Arrouays, A. Chambers, V. Chaplot, Z.-S. Chen, K. Cheng, B. S. Das, D. J. Field, A. Gimona, C. B. Hedley, S. Y. Hong, B. Mandal, B. P. Marchant, M. Martin, B. G. McConkey, V. L. Mulder, S. O'Rourke, A. C. Richer-de-Forges, I. Odeh, J. Padarian, K. Paustian, G. Pan, L. Poggio, I. Savin, V. Stolbovoy, U. Stockmann, Y. Sulaeman, C.-C. Tsui, T.-G. Vågen, B. van Wesemael and L. Winowiecki. "Soil carbon 4 per mille." *Geoderma* 292 (2017), pp59-86.

Books

Bairacli-Levy, Juliette de (1963), *Herbal Handbook for Farm and Stable*, Faber & Faber, London

Bio-dynamic Agricultural Association of Australia (BDAAA) (2023), *A Practical Introduction to the Australian Demeter Biodynamic Method of Farming*, Powelltown, Victoria

Coleby, Pat (2004), *Natural Farming*, Scribe, Brunswick, Victoria

Kevran, C. Louis (1980), *Biological Transmutations*, Happiness Press, Asheville, North Carolina

Kolisko, Lily and Kolisko, Eugen (1978), *Agriculture of Tomorrow*, Kolisko Archive, Stroud

Masson, Pierre (2014), *A Biodynamic Manual*, Floris Books, Edinburgh

Philbrick, Helen and Gregg, Richard B. (2016), *Companion Plants and How to Use Them.* Floris Books, Edinburgh

Podolinsky, Alex (1985), *Biodynamic Agriculture Introductory Lectures Volume 1*, Gavemer, St Leonards, New South Wales

Podolinsky, Alex (1989), *Biodynamic Agriculture Introductory Lectures Volume 2*, Gavemer, Sydney, New South Wales

Podolinsky, Alex (1999), *Biodynamic Agriculture Introductory Lectures Volume 3*, BDAAA, Powelltown, Victoria

Podolinsky, Alex (1990), *Active Perception*, Gavemer Foundation, Sydney, New South Wales

Podolinsky, Alex (2000), *Biodynamics Agriculture of the Future*, BDAAA, Powelltown, Victoria

Podolinsky, Alex (2000), *Living Agriculture*, BDAAA, Powelltown, Victoria

Podolinsky, Alex (2002), *Living Knowledge*, BDAAA, Powelltown, Victoria

Podolinsky, Alex (2003), *Ad Humanitatem*, BDAAA, Powelltown, Victoria

Podolinsky, Alex (2004), *FiBL Lecture*, BDAAA, Powelltown, Victoria

Podolinsky, Alex (2015), *Life Contra Burocratismus*, Alex Podolinsky, Powelltown, Victoria

Schwenk, Theodor (1965), *Sensitive Chaos*, Rudolf Steiner Press, London

Steiner, Rudolf (2004), *Agriculture Course*, Rudolf Steiner Press, Forest Row, UK

Ziegler, Arthur (2018), *Seawater Concentrate for Abundant Agriculture*, Ambrosia Technology LLC, Raymond, Washington

DVDs
Soil Conversion (Alex Podolinsky)
Soil Cultivation (Alex Podolinsky)
Stirring and Equipment (Alex Podolinsky)
Five Biodynamic Farms (Alex Podolinsky)
Cosmo-Earthly Ecology (Alex Podolinsky)
Biodynamic Composting, Farm Scale (Mark Peterson)
Biodynamic Composting, Small Scale (Rod Turner)
From Horse Paddock to Living Soil (Darren Aitken)
Living Soil to Vegetables (Darren Aitken)

Magazines
Biodynamic Growing, www.bdgrowing.com, www.biodynamicgrower.com
This magazine is a comprehensive Biodynamic resource, featuring a wide range of Biodynamic farms and gardens and including detail on most aspects of Biodynamic practice. The magazine showcases farmers and gardeners applying the Australian Demeter Biodynamic method worldwide.

All the books by Alex Podolinsky, together with the DVDs and magazine listed above, are available from www.bdgrowing.com or www.biodynamicgrower.com

Audiobooks
Growing Grain without Chemicals
www.saveoursoil.com.au

Organisations and Advisers Representing the Australian Demeter Biodynamic Method

Oceania

Australia
Bio-dynamic Agricultural Association of Australia
Training body for Biodynamic farmers.
www.demeterbiodynamic.com.au

Bio-dynamic Research Institute
Certifying body for Australian organic and Biodynamic growers.
www.demeter.org.au
Tel. +61 428 677 669
Email: info@demeter.org.au

Biodynamic Growers Australia Incorporated
Training body for small-scale growers and home gardeners.
www.demeterbiodynamic.com.au

Biodynamic Marketing Company Ltd
Australian not-for-profit wholesale distributor of Biodynamic and organic produce.
www.biodynamic.com.au

Australian Biodynamic Development Fund, Inc.
A not-for-profit organisation raising funds towards the establishment of a Biodynamic

training farm. Welcomes donations and bequests.
Email: biodynamicgrowing@gmail.com

Fix Engineering
Maker of natural farming equipment, including the Biodynamic Rehabilitator plough and seeding, weeding and crop-lifting equipment.
www.fixengineering.com.au

Asia

Pan-Asia Bio-dynamic Association
Comprising farmers in China, Taiwan, Malaysia, Thailand and Indonesia.
www.facebook.com/pabda.org/

Bio-dynamic Agricultural Association of Malaysia
www.facebook.com/p/Bio-Dynamic-Agricultural-Association-of-Malaysia-100064979071774/

France

BioDynamie Services
French organisation run by Vincent Masson, supplying Australian-style Biodynamic preparations. Also provides Biodynamic equipment, training and information.
Email: contact@biodynamie-services.fr
www.biodynamie-services.fr

Eco-Dyn
Manufacturer of Biodynamic equipment
Email: contact@ecodyn.fr
www.ecodyn.fr

Soin de la Terre– Association pour la Recherche sur les Pratiques en Agriculture BioDynamique (Care of the Earth – Association for Research on Biodynamic Practices). An association initiated by Vincent Masson, composed of scientists, intellectuals, growers and gardeners.
www.soin-de-la-terre.org

Aurélien Prouillac
Vigneron and market gardener.
Vignoble Prouillac, Sigoulès et Flaugeac, France.
Email: aurelien.prouillac@hotmail.fr

Hugues Doche
Dairy farmer in south-western France, mentored by Alex Podolinsky from 2012 to 2018. His family markets raw milk and other dairy products made on-farm, including many different cheeses (made from raw milk) and yoghurt. Biodynamic preparation-maker.
Tel.: +33 553 233 987
Email: hugo.doche@orange.fr

Italy

Agricoltura Vivente
Italian association for farmers applying the Australian Demeter Biodynamic method. Bridgette Olsen is its secretary and Biodynamic preparation supplier.
www.biodinamicapratica.it
Email: brierbette@gmail.com

Saverio Petrilli
Contact for vignerons.
Email: saveriopetrilli@yahoo.it

BioMeccanica
A company based in Reggio Emilia that builds water heaters, stirring machines (based on the Australian-developed stirrers) and sprayers for the application of Biodynamic preparations. They have agents in Europe, the US and Canada.
www.biomeccanica.com

Croatia

Udruga Stolisnik
Croatian association for the Australian Demeter Biodynamic method.
Tel.: +385 98 408777
Email: stolisnik.biodinamika@gmail.com

Denmark

Birthe Holt
Small-scale market gardener and Biodynamic preparation-maker.
www.hertha.dk
Tel.: +45 30954577
Email: birtheholt@live.dk

Many countries that currently have no organisations representing the Australian Demeter Biodynamic method do have a national Biodynamic association. Contact your national Biodynamic organisation via internet search as a starting point.

Serious professional farmers in countries lacking Australian-method preps and expertise may approach the BDAAA. The BDAAA provides training in preparation-making to suitable professional farmers who want to apply the Australian Demeter Biodynamic method, provided they are prepared to assist other farmers and supply preps in their country, to enable the method to develop in areas where it is not currently active.

About the Author

John Bradshaw was a founding member of the Biodynamic Gardeners Association (now Biodynamic Growers Australia) which began at the completion of a 1975 Biodynamic lecture series by Alex Podolinsky, and was trained by Alex and the founding leaders of the association.

He worked part time with Biodynamic dairy farmer, Henry Stephenson for one year and then farmed a 56 acre Biodynamic property in South Gippsland, Victoria, for five years until rising interest rates forced a retreat. He continued to work part-time on a variety of Biodynamic and conventional beef and dairy farms while studying and working in education. He has had a Biodynamic garden wherever he lived, supplying his family with fruit and vegetables for nearly 50 years. In 1993, he became president of the Biodynamic Gardeners Association (until 2011) and began providing Biodynamic introductory training to gardeners, smallholders and farmers, running over 50 training days to date. He wrote the first Biodynamic Practical Notes booklet, which was welcomed by Alex Podolinsky and was sent to all Australian Demeter Biodynamic method gardeners and farmers in Australia and Europe. In 2003 he launched *Biodynamic Growing* magazine which showcased the Australian Demeter Biodynamic method to a worldwide audience, featuring stories on many of Australia's and Europe's best Biodynamic farmers and gardeners, as well as on the method's increasing uptake in Asia, publishing until 2019. He returned to full-time Biodynamic farming in 2013, running a South Gippsland, Victoria property with his wife, Anna until recently, producing Demeter certified eggs, running a market garden and fattening beef cattle.

He has been a Bio-Dynamic Research Institute certification inspector and certification reviewer for over 20 years.

Index

Greenhouse gas reduction 22–23, 36, 266–267
Guano 112, 131, 199, 206, 231, 241
Gypsum 111–112

Hand stirring 500 and 501 77–78
Harrowing 7, 125–127, 195-196
Health impact of farming methods 35–36
Hens 187, 236–244
Herzeele, Albrecht von 25
History of agriculture 15–19
Horsetail 154–155, 177–178

Inoculation of legumes 130–131
Insect pests 145–152, 177–178, 207–209
Irrigation 33, 37, 195–196, 206–207, 221, 237, 256

Kolisko, Lily and Eugen 16, 21, 138–140, 144

Liebig, Justus von 16
Lice 185–187
Light 55, 83–88
Liquid manure 106–107
Lime 111
Liver fluke 187

Manure 92-95, 113–114; liquid manure 106–107
Market gardening 39, 218–226, 252–259
Mastitis 187
Micorrhizal fungi 24
Milk as anti-fungal 157
Milk fever 187
Mineral licks 182–183
Moon 8, 71; sowing 8; rhythms 138–141; and zodiac 142–144

Nodes 142

Oak bark 12
Organic agriculture 15, 19
Organic matter 3

Pasture management 7–8, 124, 194–198
Pear and cherry slug 178, 209
Peppering 148, 169–171; weeds 161
Pfeiffer, Ehrenfried 9, 16, 21
Phosphate 111
Phytophthora 154, 178
Pigs 182, 188
Pink eye 187
Planetary influences 144
Plant expression 48–52
Plant diseases 151–157, 207–209
Plant feeding, natural 2–5
Ploughing 115–123
Podolinsky, Alex 3, 10–11, 16-19, 21–24, 49, 54, 61, 219–220, 222, 268–269
Potassium 111
Poultry 187, 236–244
Prepared 500 11, 66, 70

Rehabilitator plough 118–119
Rock dust 111
Rotational grazing 7, 17, 124–126, 183, 198

Salinity 37
Scale 177; San Jose 178
Scientific studies 20–30
Seaweed 112
Seawater concentrates 113
Scaly leg 187
Scours (calves) 187
Seed development 142, 149
Seed saving 259
Sheep 181, 185–186

Other books from Hawthorn Press

Designing Regenerative Food Systems
And Why We Need Them Now
Marina O'Connell

A toolkit of six regenerative food growing systems which have been tried and tested. These can help farmers and growers transform industrial food production systems into resilient, biodiverse, carbon negative, productive farms and bring about an agroecological revolution. The book includes: Farms and garden design for growing healthy food from living soil in low input closed loop systems; The circular food economy; The four challenges of climate change mitigation, climate adaptation, offsetting biodiversity loss and producing enough healthy food for a growing population. The author's case study of her Huxhams Cross Farm researches how depleted soil was transformed into a thriving living soil drawing on the toolkit of these six sustainable methods.

Marina outlines the six main forms of sustainable food production: biodynamic, organic, permaculture, agroforestry, agroecology and regenerative farming methods. The principles and practices of each approach are explained concisely, with illustrative case studies of successful examples. There are follow up resources including articles, books, film references, and training available around the world.

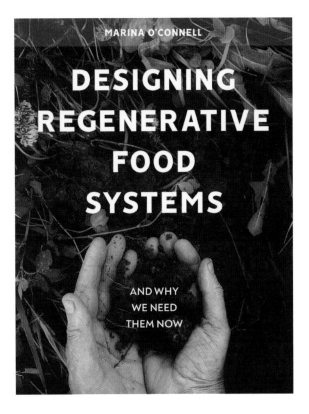

The book concludes with the Huxhams Cross Farm case study with hard research evidence, reviewing the extent to which the four challenges have been met and how a successful farm has been developed from bare land in five years. The conclusion addresses how farming can be transformed by tackling such barriers as land access for farmers, the psychology of scarcity and building farmer capacity. The key roles of food choices, policy, community supported agriculture are addressed.

224pp, 246 x 189mm; paperback; ISBN 9781912480548

Baking Real Bread
Family recipes with stories and songs for celebrating bread
Warren Lee Cohen

This is a fully revised second edition of the ever-popular *Baking Bread with Children*, first published in 2008. Share the magic of baking, with family-friendly recipes and a whole chapter on gluten-free recipes. There is advice on nutrition and wheat sensitivities, as well as tips on rebuilding the local bread economy and building a real bread culture: from farm to plate, growing healthy wheat, heritage wheats, fresh grains, home milling and supporting artisan bakeries. Fully updated and now in colour, it includes World stories, songs and poems.

224pp; 246 x 189mm; paperback; ISBN 9781912480883

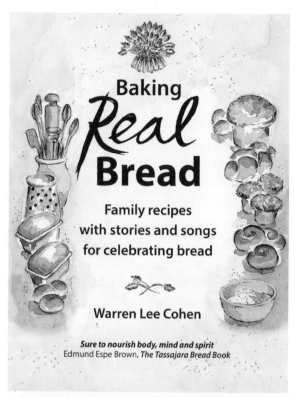

Gardening for Life
The Biodynamic Way
Maria Thun

Biodynamic techniques recognise that plant life is intimately bound up with the life of the soil; that the soil itself is alive and that the degree of vitality influences the health of the crops. You will soon be able to grow quality produce which possesses vitality and has the highest flavour, through the nurture of the soil. Whether you are an experienced gardener or not, whether or not you have used permaculture or grown organic produce before, this book offers valuable, easily accessible tips.

128pp; 210 x 160mm; paperback; ISBN 9781869890322

The Insecurity Trap
Paul Rogers & Judith Large

People ask what is going wrong with the world, with new wars, extreme populist movements, climate breakdown, poverty, inequality and exclusion. *The Insecurity Trap* addresses this sense of unease and brings together the three key challenges of the climate crisis, an unjust economic system and an approach to security rooted in hard military control, and suggests what can be done. Those concerned about climate, poverty, excessive wealth and the increased risk of war will find this short book a useful resource.

96pp; 216 x 138mm; paperback; ISBN 9781912480951

My Family and Other Allergies
Safe and scrumptious recipes for diverse diets
Julie Gritten

What can we feed our loved ones, when their needs are mixed? This book shows how you can cater for everyone at your table in a non-judging, harmonious way. Food intolerance and allergies have multiplied over recent years, with gluten allergy rising in particular. Vegetarianism and veganism have grown steadily and there are more people who choose to eat only local, sustainable, organic or raw food.

This hopeful and humorous book explains the reasons for our different food needs from ethics to the gut-brain axis. It has over 130 illustrated recipes and 20 chapters covering dietary issues such as diabetes, gluten and other intolerances, veganism, IBS and additives.

144pp; 246 x 189mm; paperback; ISBN 9781912480531

Small Steps to less Waste
Stories to Inspire Change
Claudi Williams

What are the tipping points that encourage people to make significant changes to their behaviour? Here are ten stories of personal enlightenment and the practical changes they made. Each chapter will tell the journey of each maker, showing that people of all ages can develop the skills and courage to make, create and look after what they have, rather than throw away and buy new. With tips on how to 'repair, reuse, recycle' and why, this book shows how to save money, cut waste and reduce consumption, offering simple alternatives to mass-produced, shop-bought, highly packaged goods.

96pp; 256 x 186; paperback; ISBN 9781912480296

Ordering Books

If you have difficulties ordering Hawthorn Press books from a bookshop, you can order direct from our website www.hawthornpress.com, or from our UK distributor BookSource:

50 Cambuslang Road, Glasgow, G32 8NBTel: (0845) 370 0063,

Email: orders@booksource.net. Details of our overseas distributors can be found on our website.

Hawthorn Press

www.hawthornpress.com